CAMBRIDGE COUNTY GEOGRAPHIES

SCOTLAND

General Editor : W. Murison, M.A.

T0352320

CLACKMANNAN

AND

KINROSS

Cambridge County Geographies

CLACKMANNAN

AND

KINROSS

by

J. P. DAY, B.A., B.Sc.

Fellow of the Royal Geographical Society ;
Fellow of the Royal Scottish Geographical Society ;
Examiner in Geography in the University of London

With Maps, Diagrams and Illustrations

Cambridge :

at the University Press

1915

CAMBRIDGE UNIVERSITY PRESS
Cambridge, New York, Melbourne, Madrid, Cape Town,
Singapore, São Paulo, Delhi, Mexico City

Cambridge University Press
The Edinburgh Building, Cambridge CB2 8RU, UK

Published in the United States of America by Cambridge University Press, New York

www.cambridge.org
Information on this title: www.cambridge.org/9781107659391

First published 1915
First paperback edition 2013

A catalogue record for this publication is available from the British Library

ISBN 978-1-107-65939-1 Paperback

This publication reproduces the text of the original edition of the
Cambridge County Geographies. The content of this publication has
not been updated. Cambridge University Press has no responsibility
for the accuracy of the geographical guidance or other information
contained in this publication, and does not guarantee that such
content is, or will remain, accurate.

CONTENTS

CLACKMANNANSHIRE

CONTENTS

KINROSS-SHIRE

ILLUSTRATIONS

MAPS AND DIAGRAMS

The illustrations on pp. 89, 97, 105, 111, 115, 124 are from
photographs by Mr A. C. M'Nair; those on pp. 4, 9, 12, 17, 22,
26, 33, 38, 40, 44, 46, 49, 53, 57, 59, 65, 67, 78, 90, 99, 108,
116, 125, 127, 129, 132, 137 and 140 are from photographs by
Messrs J. Valentine & Sons; the portraits on pp. 70, 72 and
136 are from photographs by Messrs T. R. Annan & Sons; those
on pp. 86 and 95 are from the *Guide to Kinross-shire* published by
the *Kinross-shire Advertiser*; those on pp. 119 and 131 are re-
produced by kind permission of the Society of Antiquaries of
Scotland.

The author wishes to acknowledge gratefully help from
Dr William Smith and from several other gentlemen both within
and without the counties; he is also indebted to the well-known
work of Messrs Macgibbon & Ross for guidance as to architectural
features of interest.

CLACKMANNANSHIRE

1. County and Shire. The Origin of Clackmannanshire.

For administrative purposes Great Britain is now divided into a number of separate areas known as counties, whose boundaries are based upon older territorial divisions. The older divisions were, in many cases, known as shires, a name signifying a portion of land under the control of some distinct authority, and derived from Anglo-Saxon *scir*, office, charge, administration. The historical development of the modern county from the ancient shire has not, however, been uniform ; there are counties which were never called shires, and there were shires which have not become counties, such, for example, as the Bishopshire, a district almost coincident with the present parish of Portmoak, in Kinross-shire, which was formerly church land under the control of the Bishop of St Andrews. Generally speaking, however, the Anglo-Saxon shire came to mean the district under the rule of an ealdorman or earl, a title superseded later by *comes* or count; hence the name *county*. Within this district there was a permanent direct representative of the King : the *vice-comes*, shire-reeve or sheriff. In England the course of constitutional development tended to increase the power of the central authority

and, from the time of William I, the earl or count become
a titular grandee rather than a viceroy, the actual power
passing to the royal officer, the sheriff, and later to the
justices.   In Scotland the development was different ; the
royal power was not strong enough to prevent the great
feudal lords from capturing the office of sheriff, which, as
in Gascony, became in many cases hereditary in certain
families till the abolition of all heritable jurisdictions in
1748.

Between the Firth of Tay and the Firth of Forth,
and separated on the west from the rest of Scotland by
the high barrier of the Ochils, lies the peninsula of Fife.
Though distinctly one natural geographical region, this
peninsula contains three counties : Fife, occupying almost
four-fifths of the whole ; and, occupying the remaining
portion, Clackmannan and Kinross, the two smallest
counties of Scotland.   When we remember that the
counties of Scotland are administrative areas formed by
historical development upon the basis of old territorial
divisions, it seems strange that this clearly-defined peninsula, anciently known by the one name of Ross, should
have become subdivided into three counties ; nor is it
easy to discover at what period Clackmannan first came
to possess the status of an independent county.   The area
styled Clackmannanshire has varied in extent from time
to time ; while, for administrative and judicial purposes,
the shire has been united first to one and then to another
of the adjacent counties.   For parliamentary representation
it is united with the shire of Kinross ; but, as regards the
Sheriff's jurisdiction, though, curiously enough, Clack-

mannan never appears in recent times to have been united
with Fife, we find it joined with Kinross in 1807, with
Linlithgow in 1853, and, since 1870, with Dumbarton
and Stirling. Such frequent changes of jurisdiction are
among the disadvantages which result when the administrative area is not coincident with a natural region ;
and this same cause has led to the alterations in the
county boundaries.

Their most recent readjustment was made by the
Boundary Commissioners in 1891. Previously the county
contained four parishes : Alloa, Clackmannan, Dollar and
Tillicoultry, and portions of the parishes of Logie and
Stirling. The Clackmannan portion of Stirling parish
was transferred to Stirlingshire. The Clackmannan
portion of Logie parish was divided into three parts. One
was transferred to the Stirlingshire parish of Logie ; a
second was united with the parish of Alva, which was
wholly given to Clackmannanshire ; while the third part
was absorbed by the parish of Alloa.

The county takes its name from the small town of
Clackmannan, which, for this reason, has some claim to
be considered the county town, though Alloa, with a
population nine times as large, is the chief burgh and the
administrative centre. The name Clackmannan is of
Gaelic origin. *Clachan* signifies the stones, and, being
frequently used of the stones which mark a burial ground,
it came to signify the church, and, finally, the kirkton or
village. Clackmannan or Clachan-Mannan is generally
accepted as meaning the stone circle, or village, of the
ancient district called Manann, which lay at the head of

Cross and Main Street, Clackmannan

the Forth estuary. Stirling and the valleys focussed at Stirling were the essential parts of the district, which extended towards the hills to boundaries, not precisely known and probably always indefinite. Slamannan—signifying the moor of Manann—is the name of a village and parish some four miles south of Falkirk. What may be termed the "nodality" of Stirling made it more and more the geographical centre of the district, and so the greater part of Manann came to be called Stirlingshire.

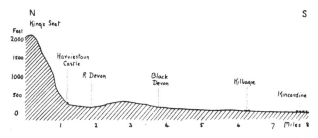

Section from King's Seat Hill to Kincardine-on-Forth

That part, however, which stretches between the Ochils and the Forth towards Fife was to a certain extent more secluded than the other radial carses, its western entry being blocked by the Abbey Craig. In the saddle between the Abbey Craig and the Ochils lay forests, marshes and lakes ; between the steep southern face of the Craig and the river lay but a narrow band of undrained marshy ground. Shut off by such obstacles from easy communication with the rest of Manann, Clackmannan never became included in Stirlingshire. That it also remained

excluded from Fife is perhaps due to somewhat similar
obstacles on its eastern border. From the then marshy
valley of the Devon below King's Seat Hill to Longannet
Point on the Forth, there formerly stretched a belt of
alternating woodland and marsh, whose position is indicated
by Tulliallan Forest and Bogside in the Fifeshire parish
of Tulliallan, and by The Forest in the parish of Clackmannan. This small area, shut in by the Ochils, the
Forth, the forested land and the Abbey Craig, came to be
known as Clackmannanshire. But these natural boundaries no longer delimit the county to-day. The marshes
have been drained and the natural forests almost entirely
cut down ; the western boundary no longer reaches to
the Abbey Craig ; while the northern boundary extends
so far beyond the edge of the plain as to include a hilly
upland area almost as large as that of the low-lying ground.
Only on the southern side is the ancient natural boundary
yet maintained, the county still extending to, but not
passing beyond, the river Forth.

2. Position. General Characteristics. Size. Boundaries.

Clackmannanshire is the southern threshold of the
Fife peninsula as Newburgh is the northern. The Ochil
range, rising to over 700 feet within a mile of the Tay
at Newburgh, runs across the base of the peninsula to
end abruptly in the lofty hills which overlook the valley
of the lower Devon from Dollar to Menstrie. This

continuous band of uplands presents a barrier to communication between the Fife peninsula and the rest of Scotland.
It is true that the Ochils are traversed near their centre
by a road which, passing up Glen Devon, crosses the
water-parting at a height of 750 feet above sea-level and
follows the narrow and deep trench of Glen Eagles to
Strathearn ; but most of the traffic prefers the easier routes
into the peninsula, on the north side by Glen Farg or
Newburgh, on the south side through Clackmannanshire.
The first place up the river Forth at which a bridge for
foot passengers has been built is at Stirling, and there,
from all parts of the lowlands of Scotland, are focussed
roads which pass eastwards between the steep edge of the
Ochils and the Forth along the narrow plain of Clackmannan into Fife.

This position of Clackmannanshire as the southern
gate of Fife has had a persistent influence on the importance and prosperity of the county ; but the influence has
tended to diminish.   An alternative means of entry into
Fife from the south is the ferry across the Forth ; and
from the earliest times much of the traffic, particularly
the passenger traffic from the capital, has gone by the
Queensferry passage.   The expense and inconvenience,
necessarily attending the change from land to water
carriage, and the often stormy character of the Firth,
would operate to preserve for the longer route by Stirling
and Clackmannan a considerable share of the southern
traffic, but, with the development of a really efficient and
reliable ferry service, this share has become less.   To-day
a regular ferry service is maintained at four points : from

Granton to Burntisland ; between the Queensferries ; at Kincardine ; and at Alloa. A more serious competitor arose with the construction of the railway bridges. The traffic from the south now passes into Fife mainly by the Forth Bridge, between which and Stirling another railway, from Glasgow, crosses the river to the south of Alloa. Thus, though the southern traffic with Fife has been largely diverted from the route through Clackmannanshire, the stream of the important western traffic with the peninsula still passes along the narrow gateway between the Ochils and the Forth.

The position of Clackmannanshire as the southern threshold of Fife is, therefore, a geographical fact of fundamental importance and one which throws light upon the past history of the county and the present economic conditions. The most striking episodes in the history are connected not so much with the deeds of the inhabitants as with those of travellers journeying across the county ; legends of the passing of St Serf or stories of the fiercer transit of Montrose. Again, the characteristic feature of the shire as a thoroughfare has influenced the distribution of the towns and villages, strung out as they all are in one or other of two long lines from west to east : Menstrie, Alva, Tillicoultry, and Dollar on the road to Kinross ; Tullibody, Alloa, and Clackmannan on the road to Dunfermline (see map on p. 63).

But the importance and prosperity of Clackmannanshire are by no means entirely due to its advantageous position as the southern gate of the peninsula. In itself, the county possesses resources more considerable in value and much greater in variety than its size would lead one to expect. Indeed the variety of its sources of wealth is a very noticeable characteristic. Within its northern confines more than fifteen thousand sheep pasture on the grass-covered slopes of the Ochils; the rich alluvial soils of the Devon and Forth valleys are valuable agricultural

Stirling Street, Menstrie

land; the coalfields, which underlie the southern half of the county, provide work for more than thirteen hundred miners and supply a basis for the export trade of Alloa; more than half the occupied women in the county are engaged in the prosperous woollen industry; and minor industries, such as brewing and distilling, brick, cement, glass, and pottery works, are not lacking. Mining, manufacturing, agriculture, stock-raising, and commerce, all contribute to make Clackmannanshire, in proportion to its size, one of the wealthiest counties of Scotland.

The total area of Clackmannanshire is 35,214 acres, or, excluding water, 34,927 acres. It is, indeed, by far the smallest county in the British Isles, as may be seen by a comparison with the smallest counties of England, Wales, and Ireland—London, 74,817 acres; Flint, 163,025 acres and Carlow 221,424 acres. The average size of a Scottish county is 577,893 acres, or more than sixteen times the size of Clackmannanshire. The greatest length of the county from north to south is nine miles, its greatest breadth from west to east is nearly ten. In shape the shire is compact, and a circle of 4½ miles radius drawn from Tillicoultry as centre would include practically the whole area. Least of all lands though it is, the men of Clackmannan are proud of their home, and wherever counties compete—whether in cricket or in more serious arenas—have so acquitted themselves as to win for the shire the epithet of "gallant."

On the south-west side the county is bounded for nearly eleven miles by the windings of the river Forth, and this is the only considerable stretch where the modern boundary is a natural one; elsewhere it is erratic, following neither river valley nor water-parting for more than a mile or so at a time, so that its course is more easily followed on the map than described in words. The north of Clackmannan is bounded by Perthshire, the east by Kinross and Fife, and the west by Stirlingshire. Two islands, Tullibody Inch and Alloa Inch, in the Forth have been detached from the mainland by the meanderings of the river. Both are included in Clackmannanshire, but only the latter is inhabited, its population at the time of the 1911 Census being returned as three males and two females.

3. Geology.

From the rocks we may learn of changes of the physical conditions, fluctuations in the climate, and alterations in the elevation, which took place long ages before the advent of man. Enclosed within the rocks of some mountain top are found the fossilised remains of marine creatures, and we know that the land there, now thousands of feet above the sea-level, was once many fathoms beneath it. In the coal seams of a deep mine are seen the decayed and compressed remains of vegetation which once throve exposed to the light and air of the surface. Embedded in a slab of clay is the imprint of a tropical palm, mute witness to a time when these islands experienced a torrid climate; while from the ice-smoothed rocks of the hill-side we learn that at another period a vast ice-sheet overrode the country. Stupendous changes are these, from ocean depth to mountain top, from tropical conditions to arctic—changes which can be understood aright only when it is remembered that many thousands of years went to every stage of the long history. Not by any violent cataclysm were the former mountains buried, nor was the ocean bed raised by any sudden upheaval;

In Alva Glen

but the alterations in elevation and the fluctuations of climate were perhaps as slow and imperceptible then as the changes now taking place are to us. It is part of the work of the geologist to attempt to trace the history of these secular changes; and, for the accomplishment of this, he relies partly upon the position and character of the rocks now exposed, but mainly upon the kind of fossils they contain. From such evidence the world's organic history has been divided into three main stages. The first great period is called the Palaeozoic era, because it includes the earliest forms of life; the last period is the Cainozoic, with the recent forms of life; while between these two is the Mesozoic or middle era, though in point of duration the Palaeozoic era was longer than the two later eras together. Of course, some of the Palaeozoic forms are found also in Mesozoic rocks, and Mesozoic forms may continue into Cainozoic times; but yet these three divisions are separated by periods of change relatively so rapid as to cause a sufficiently distinctive difference in the general characteristics of the organic remains of each division. The Mesozoic era, for example, is especially distinguished by the enormous and wonderful reptiles then existing on the land and in the sea, so that it is sometimes called the Age of Reptiles. These reptilian forms are not preserved except in rocks of Mesozoic age. But it is the Palaeozoic period that concerns us in studying Clackmannan, for the rocks now underlying the soil of the county were originally laid down in that era. We can narrow the age of the different rocks of the Palaeozoic period by subdividing them into groups, named usually from the area where they were first studied or are best developed, like the Cambrian rocks of Wales, but sometimes from the character of the rock itself, like the Old Red Sandstone. The following table gives the succession of the Palaeozoic rocks, the oldest formation being at the bottom of the list :

> Permian.
> Carboniferous.
> Old Red Sandstone or Devonian.
> Silurian.
> Ordovician.
> Cambrian.

The study of the rocks is of interest not merely as enabling us to follow the development of organic history, but also because the character of the rocks now at the surface, their structure and position, have determined the present-day physical features of the land. The position and form of the hills and mountains of Scotland are not directly due to any uplifting of certain portions of the earth's crust, but are the result of the continuous attacks of rain, rivers, frost, ice, and wind. All these agents tend, in varying degrees of power, to wear down the land surface ; and, since some rocks are more easily removed, either by disintegration or dissolution, than others, the more resistant rocks remain outstanding as high ground. Now the stratified rocks, those which have been laid down under water as a sediment in layers, are, generally speaking, more easily disintegrated than the massive igneous rocks, those which have been erupted from the heated interior of the earth. Hence, in this part of the country, the igneous rocks are left as hills overlooking the plains of the less-resistant sedimentary deposits.

We are now in a position to attempt to trace in broad outline the physical history of Clackmannan. The period when, long ages ago, the Old Red Sandstone of the midland valley of Scotland was being deposited in some ancient inland sea or lake, was a time of volcanic activity. Through the floor of that sheet of water great masses of lava and volcanic conglomerate were ejected to spread over the sandstone and to be in turn themselves buried beneath later deposits. After numerous and varied changes the overlying deposits were removed by denudation, until to-day the volcanic masses are exposed and form long lines of uplands, such as the Sidlaws, the Pentlands, and the Ochils. The south-west extremity of the Ochils occupies the northern half of Clackmannanshire ; and the whole character of this northern half is in marked contrast to the low-lying Carboniferous plain between the hill foot and the Forth—differing most strikingly in elevation, in scenery, in economic value, and in the density of habitation. The difference is the more noticeable since the igneous rocks do not pass gradually under the Carboniferous ; instead they come to an abrupt end along a line immediately to the north of the Menstrie-to-Dollar road. This important feature requires further explanation. The comparatively stiff outer crust of the Earth is not free to shrink in the same way as the rocks, presumably in a molten state, of the viscid interior. It can accommodate itself to a smaller circumference only by being squeezed into ridges and troughs, or by fracture and dislocation; and the attempt so to accommodate itself results both in slow secular upheaval and in earthquake and fracture. The Ochils themselves have been squeezed up into a broad arch, while examples of fracture and dislocation may be found in almost any district. Such fractures, called by geologists faults, may occur without displacement of the rocks on either side; but, in general, the rocks slide along the plane of division. The length of the slide, or the "throw" as it is called, may vary from a few inches to many thousands of feet. Again, the horizontal extent of the fault may run for a few feet only or stretch for hundreds of miles. One great fault begins near Stirling and may be traced in a line across Clackmannan a little to the north of the Menstrie-to-Dollar road. Beyond Dollar it appears to split into two branches. It is this fault that causes the igneous rocks of the Ochils to come to an abrupt end in a great wall; and the rocks on the southern side are estimated to have been displaced to a depth of at least 10,000 feet. One might readily suppose that this enormous displacement was the direct cause of the present difference in elevation; in other words, that the step-like structure resulting from the throw has persisted till to-day, but this has not been the case. Both the igneous rocks of the present hills and the carboniferous rocks of the modern plain were buried under hundreds, nay, thousands, of feet of later sedimentary deposits. These deposits have been gradually removed by the action of a stupendous denudation continued through long ages, and it is only because the Ochils are

The Ochils at Alva

composed of the harder, more resistant igneous rocks that they stand to-day as a great northern wall overlooking and sheltering the worn-down plain of the more easily removable Carboniferous rocks. Nevertheless, we must remember that, but for the fault, the igneous rocks, and therefore the Ochils, would be continued further southwards than they are, and neither the low plain of Clackmannan nor its coalfield would be occupying their present position. This consideration gives weight to the contention that the great fracture was the most important incident in the physical history of the shire.

Another important event must also be noted. In the latter part of the Cainozoic era, at a time which seems to have immediately preceded the advent of man, the climate of these latitudes was intensely cold. Britain, as far south as the Thames, was at one period covered by a thick sheet of ice just as Greenland is now. Before the advance of this ice-sheet from the north, the physical features of the country were, in general outline, much as they are to-day ; but the land then stood somewhat higher and had been sculptured by a process of long continued denudation. The hills had been carved into sharp ridges, bold peaks and rugged upper slopes, while the valleys and plains were covered with a thick blanket of residual waste. As the arctic conditions came on, the ice-sheet, gathering among the high hills of the north-west, pushed slowly southwards, sending long tongues of ice down the valleys, lapping round the hills, until at length it so increased in height as to override the whole countryside. As the heavy mass slowly ground its way along, it carried before it the loose waste material of the valleys and plains, and, plucking away the rugged projections of the crests and sides of the hills, scoured and ground the bare rock to a smooth and polished surface. This period is called the Glacial Epoch or Ice Age. Eventually, with a return of more genial conditions, the southern portions of the great glaciers melted and the ice-sheet seemed to retreat northwards, though both the advance and retreat were intermittent rather than continuous. When the ice-sheet melted away, it left behind the masses of ground-up material which it had been transporting. This waste now covers the Carboniferous rocks of Clackmannan, and can be traced up the valleys of the Ochils, sometimes more than 1000 feet above sea-level. It varies in texture, composition and colour according to its origin. Most often it is a stiff clay full of boulders—hence its name of Boulder Clay—and fragmentary material, and sometimes containing lime, a soluble mineral seldom found in soils produced mainly by weathering. The clay is dark blue if derived from carbonaceous rocks ; but, if derived from the Old Red Sandstone, it is reddish, and loose and sandy. These sheets of clay are thickest on the lee side of the hills, that is, the side opposite to that from which the ice-sheet advanced. In places the Clay has been covered by deposits of sand and gravel, laid down, as their water-worn material evidences, by streams whose source may have been the retreating glacier.

Since the time of the Glacial Epoch, the land surface has been attacked by the ordinary agents of denudation ; frost has been the means of loosening the exposed rocks ;

ground water, percolating through the porous strata, has washed away soluble minerals; rivers, cutting trenches through the boulder clay, have worked over and sorted the material of their beds; but the time which has elapsed has not been long enough to obliterate the work of the overriding ice. Wherever superficial deposits have till now covered them, rock surfaces scratched by angular fragments embedded in the base of the long-vanished moving ice may be found here and there among the higher ground of the Ochils. The direction in which such markings, or striae, run, shows that the ice moved over the hills in a general east-south-east course, trending more to the east as the lower ground was reached. Great boulders, called erratic blocks because they are quite different in character from the rocks on which they now rest, lie scattered over the area to which they were borne on the surface of the ice-sheet. One feature, however, of the landscape left by the Ice Age has almost vanished. The surface characteristic of a glaciated country is an irregular one; on the higher ground, ice-scooped basins are found; on the lower, depressions of various sizes and shapes amongst the lines of sand and gravel drift. These hollows became filled with water, but, of the numerous small lakes so formed, less than a dozen remain in the peninsula to-day, though several others appear in our older maps. Streams flowing into them have helped to fill them up; the marginal marshy vegetation has crept in, and, dying, formed a peaty bottom, over which the advance has been continued. In time the sheet of water becomes a marsh, the marsh a peat-bog. With the need for more arable land, the peat-bog is drained ; and, perhaps after the removal of the peat, the land is ploughed ; then nothing remains to mark the site of the lake but the black soil upturned by the plough, or, where the land is not under cultivation, the character of the vegetation.   Occasionally too, such names as Myreside or Bogside help us to fix the site of the vanished water.   The only lake in Clackmannan to-day is Gartmorn Dam, but this, as the name suggests, is of artificial origin.   It was formed in 1700 by damming back the upper waters of a tributary of the Black Devon ; and, repaired and improved in later years, notably in 1827 and 1867, it now supplies water to the burgh of Alloa.

4. Surface Features.

From the account of the geology of the shire, we are prepared to find that the surface features north of the great fault differ absolutely from those to the south.   To the north the land has an average elevation of more than 1500 feet, and reaches its culminating point in Bencleuch, the "stony mountain," a peak, 2363 feet above sea-level, occupying the centre of the Clackmannanshire Ochils. From Bencleuch the crest line runs west by north to Blairdenon Hill (2072 feet), east to Whitewisp Hill (2110 feet), and east by south to King's Seat Hill (2111 feet).   On either side of this divide, small streams, swollen into torrents after heavy rainfall, tumble rapidly down to join the larger rivers of the lower valleys.   Curiously enough, they all

join the same river, because the Devon, which flows
eastward on the north side of this divide and roughly
parallel to it, makes a great bend beyond the eastern
confines of the county and thus returns to pass westwards
close along the hill foot across the breadth of Clackmannanshire, receiving in its passage the waters of the streams
which drain the southern face of the Ochils. These

Bencleuch

streams have cut deep and picturesque glens amongst the
hills ; and where the glens debouch upon the Carboniferous
plain the towns on the Kinross road have arisen. Thus
the Burn of Sorrow, flowing from Maddy Moss in the
basin between Whitewisp and King's Seat Hills, after
being joined by the Burn of Care at Castle Campbell,
enters the Devon at Dollar; the Dai Glen Burn, rein-
forced by the Gannel or Gloomingside Burn, enters at

Tillicoultry ; the Alva Burn at Alva ; the Menstrie
Burn at Menstrie.   Blairdenon Hill, right on the county
boundary, is an important hydrographic centre, streams
radiating from it in all directions.   The hills themselves
are covered with a thin soil formed of the detritus of the
underlying igneous rock, which, indeed, is not always
completely covered but in places protrudes in rounded
knolls or bluff crags.

South of the great fault is all low flat agricultural
country.   This area, triangular in shape with the apex
pointing west, may be divided into three parts.   In the
north, occupying a belt of ground immediately below the
Ochil edge, is the Devon valley stretching completely
across the county.   In the south, running parallel to the
Devon valley, is that of the Black Devon.   Between
these valleys, a wedge of land of slightly higher elevation
is pushed westwards from the Cleish Hills in Fife and,
some 400 feet high when it crosses the county boundary,
sinks gradually down to the general level of the plain.
Generally speaking, sandy and gravelly drift covers the
rocks of the Carboniferous plain, giving rise to a somewhat
light, loamy soil.

5. Rivers.

A navigable part of the Forth forms the south-western
boundary of the county.   At the western border of
Clackmannan, it has a breadth of about one furlong,
which gradually increases as it pursues its winding course
eastwards.   At Cambus the volume of the river receives

an addition from the Devon. Then the Forth takes a large bend to the south, flowing round Tullibody Inch and round Alloa Inch to the port of Alloa. The sinuous course of the river is due to the flatness of the plain over which it wanders. Where the gradient of a river bed is but slight, any small obstruction deflects its course. The river flows round the obstruction and the current on the outside of the loop is naturally swifter than that of the sluggish water edging, so to speak, round the inside. The faster the current, the greater is its carrying capacity and the more powerful its erosive force. The bank on the outside of the loop gets worn away while, on the inner side, the checked current will deposit some of the heavier material it was transporting. Thus the loop becomes more and more pronounced, until perhaps its narrowed neck is broken down in some time of flood and the river once more straightens its course, the detached portion of land forming an island in the stream. Later, the river ceases to use the old loop, which changes from river bed to marsh, and eventually to a mere horseshoe-shaped depression. The island now becomes united to the opposite bank as if it had migrated from one side to the other. From Stirling to Alloa in a straight line is but slightly over 5½ miles, yet the distance by the winding river is seven miles longer. Such winding loops are usually called "meanders," but here they are known as the Links of the Forth. The great agricultural value of the alluvial land originated the popular jingle:

> "A crook of the Forth
> Is worth an earldom of the North."

Alloa itself is situated on the outside of a bend and thus has the advantage of the swifter and, therefore, deeper water, though, even so, the harbour suffers from a continual deposition of sediment.   Below Alloa, the Forth, ceasing to wind, flows in a fairly straight south-east course past the two small harbours of Clackmannan Pow, at the mouth of the Black Devon, and Kennetpans, until, just after passing the eastern boundary of the county, where the river's breadth is but seven furlongs, it opens out into the estuary.

The next largest river is the Devon, rising in Blairdenon Hill at an altitude of some 1800 feet.   Then it flows, with a bend towards the north, eastwards through Perthshire territory for fourteen miles to the Crook of Devon.   There making an abrupt turn, it pursues a west-south-west course through a deep ravine to enter Clackmannanshire again at Dollar.   This stretch is most famous for its picturesque and wild scenery.   Foaming and rushing along at the bottom of a deep and, in places, overhanging gorge, the river leaps in hidden waterfalls into such gloomy chasms as the Devil's Mill, whence the tormented and tossing water passes onward beneath the Rumbling Bridge to career wildly in alternate leaps and swirling rushes down the Caldron Linn.   With the river's re-entry into Clackmannanshire, the character of the valley changes, the steep descent of the narrow gorge is succeeded by the gentle gradient of the fertile plain, and the river winds placidly westward through a broad belt of pleasant agricultural land under the shelter of the bold outline of the Ochils.   It is this green valley, "where

Devon, sweet Devon, meandering flows," that was familiar to Burns, who, on his visit to Harviestoun in 1787, wrote the short lyric commencing :

"How pleasant the banks of the clear winding Devon,
  With green spreading bushes, and flowers blooming fair!"

The Devon, above Vicar's Bridge, Dollar

On reaching the western border of the county, the river turns sharply southwards to join the Forth at Cambus, after a course of nearly 34 miles, though in a direct line Cambus is but $5\frac{1}{4}$ miles from the river's source.

Thus the Devon has two distinct sections: an upper course flowing generally eastwards to the Crook of Devon, and a lower flowing westwards from the Crook to Cambus.

This lower section, when seen on the map, appears strikingly peculiar, since the river flows in an opposite direction to that followed by the Forth, entering that river at quite an unusual angle. This curious feature is intelligible when we consider the effect of the great Ochil fault on the development of the river system. We may conceive the original Devon as a small tributary coming down from the southern face of the Ochils and receiving on its way to the Forth additional streams from the eastern slopes of the Abbey Craig. This little, but rapid, stream would gradually cut its bed deeper and deeper and would develop side tributaries from the east. Similar streams from the Ochils would run down towards the Forth, tending always to follow the easiest route. Now, the line of fracture of the fault is a line of weakness, along which the streams could most easily excavate a bed. The tributary, then, which followed the line of the fault would rapidly erode its bed ; the land at its source, coming thus to have a steeper slope, would weather down at a comparatively rapid rate, and the valley would, so to speak, eat its way up stream. Progressing in this manner, it might tap the head waters of other streams draining the Ochil face, and the river, increasing thus in volume and erosive power, would continue the process by which the divide or water-parting gets pushed further and further back until it not only has captured the streams draining the southern face of the Ochils but has also tapped the head-waters of a river running in the transverse valley behind the Blairdenon-to-Whitewisp ridge. This latter river, now represented by the upper course of the Devon,

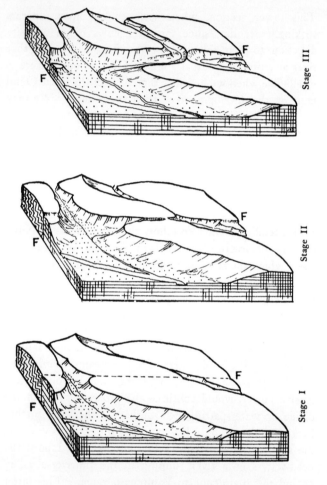

Diagrams illustrating river "piracy"
FF a line of weakness

then flowed straight on past the Crook of Devon along a
course corresponding to the Gairney Water into the plain
of Kinross, whence it issued to the sea either by the
Leven or the Eden. Once the upper waters of the
Devon were captured at the Crook and turned away
south-west, the former continuation of the upper stream,
robbed of its old supply, dwindled into the insignificant
brook now called the Gairney Water.

Another feature of interest in the lower Devon valley
is that the present alluvium hides a deep buried channel.
Near Tillicoultry a bore struck solid rock at a depth of
342 feet ; another near Alva at 336 feet ; a third near
Menstrie at 372 feet. These borings show that the old
river channel is now about 300 feet below sea-level ; and,
therefore, the land must have formerly stood higher.
With the gradual sinking of the land the rate of the swift
and powerful river would become lessened ; and so the
gorge would be gradually filled up by the sediment deposited by the stream.

As a fishing stream, the Devon has a good reputation
for trout, which average rather less than half a pound
each ; but the sport is not so good now as formerly, when
trout and parr were abundant, with pike and eels in the
deep pools and, in the spawning season, salmon from the
Forth.

The third river of Clackmannanshire, the Black
Devon, rises in the Cleish hills and flows in a course
generally parallel to the Devon, along the north border
of The Forest, by the town of Clackmannan to enter the
Forth at Clackmannan Pow. With an average gradient

of 1 in 134, it is not a rapid stream and seems to take its name from the contrast between its dull waters and those of the "crystal" Devon.

6. Natural History.

The character of the vegetation of any district depends for the most part on climate, soil, and human activity. Clackmannanshire, therefore, has two quite different zones of vegetation: the cultivated land of the warmer and drier Carboniferous plain; and the hill pasture of the colder and wetter igneous Ochils. Practically the whole of the plain is cultivated, being under crops, grass, or plantations. Little wheat or barley is grown, the chief cereal being oats. A feature of interest is the existence of several meadows sown with Timothy grass (*Phleum pratense*), for which the coal mines provide a good local market. The plantations, though few of them are more than a century old, cover in the aggregate a considerable area and, stretching on either hand above the flat meadow-lands of the Devon or scattered over the higher central ground of the plain, form a notable feature of the land-scape. In Tullibody woods the Scots pine (*Pinus sylvestris*) is dominant on the higher ground, and associated with it is spruce (*Picea excelsa*). The lower ground is occupied by oak, ash, beech, horse-chestnut, plane, and poplar. The higher wood on Wood Hill between Alva and Tillicoultry is mentioned by the late Robert Smith as being planted with mixed trees; beech, oak and Scots pine predominating and flourishing well, while less abundant are ash, elm, larch, spruce, birch, and rowan. The "Forest" at Gartlove is mainly coniferous (Scots pine and spruce).

The Clackmannanshire Ochils are clothed with grass to the summits. Two zones of vegetation may be distinguished:

A. Above the cultivated area is a zone of mixed grasses, with well marked presence of bracken (up to 1200 feet on Middle Hill, to 1000 feet on King's Seat) and whin (up to about 500 feet). Juniper also occurs occasionally in the bracken zone. This mixed herbage provides excellent pasturage for sheep.

B. Above 1000 feet or so, zone A passes into the summit *Nardus* (white moor grass) zone, which is a strong feature of the Ochils. This *Nardus* zone carries a much less varied herbage, as *Nardus stricta*, *Molinia* (blow grass) and *Agrostis* (moor-bent) occupy practically all the ground. It is a very inferior pasturage except for the June-July grazing, when sheep may be seen on the very summits, *e.g.* at King's Seat, 2111 feet high. The upper *Nardus* zone was possibly once heather land; blaeberry and heather, though almost completely suppressed by the long-continued use of the hills for pasturage, occur here and there in a stunted condition. Arctic-alpine plants are very rare on the Ochils: the hills are too rounded, and coated with glacial deposits; they have few easily weathered crags to form shelves and crannies which suit some species; nor have they alpine peat suitable to others. The following have been recorded, though in some cases the record is doubtful: *Alchemilla alpina*,

Carex rigida, *Epilobium alpinum*, *Gnaphalium supinum*, *Meum athamanticum*, *Poa alpina*, *Polygonum viviparum*, *Rubus chamaemorus*, *Salix lapponum*, *Saxifraga stellaris*, *S. hypnoides*, *S. hirculus*, *Silene acaulis*. *Saxifraga hirculus* is an extremely rare plant in any country ; it was reported as being discovered in Maddy Moss, north of King's Seat.

The lower parts of the Ochil glens are occupied by deciduous woods, while bracken flourishes on the slopes. The abrupt steep gorges of these glens offer many habitats for different species ; the glens, being sheltered and having a constant water supply, are consequently rich in plant life. For the Silver Glen at Alva, an unpublished list of Robert Smith's records 24 flowering plants and ferns on the rock ledges, and 29 along the stream margins. These lists were compiled in September and do not include any very rare plants.

Clackmannanshire is not peculiar in its fauna. The usual common wild animals are found, such as the rabbit, hedgehog, mole, squirrel and brown rat. The roe-deer, formerly to be found in Tullibody woods, is not now met with in the county. The otter was thought to have become extinct, but still makes a rare appearance in the Devon river ; one was captured as recently as May, 1913. Of birds, several, such as the bittern, the heron, and the snipe, have vanished with the draining of the bogs and marshes. The glede has also disappeared and the goldfinch is very rarely seen.

Craighorn Fall, Alva

7. Climate.

The chief factors which determine climate are latitude, proximity to the ocean, the direction of the prevalent winds, and the elevation and exposure of the land. Considering Scotland as a whole, we find that the fairly high latitude (55°–59° N.) gives a moderately cool climate with a mean annual temperature of 46·4° F. ; that the proximity of the Atlantic Ocean and the fact that the prevalent winds are from the south-west are responsible for the moist equable climate, the mean annual range of temperature being less than 20° F. The west side, facing the Atlantic, has the smaller range of temperature and the heavier rainfall. The rain comes at all seasons of the year but especially during the winter months, the mean annual amount for Scotland being 40·02 inches. Local variations in climate are chiefly due to differences in elevation and exposure.

Clackmannanshire, being so small, does not present any marked differences in its climate from that of the east of Scotland generally ; moreover, since there are not any stations in the county sending in returns to the Meteorological Office, it is difficult to obtain reliable statistics of the actual climate experienced. The tables published from the Reports of that office by the Board of Agriculture divide Scotland into four divisions. Now the mean January temperature for the north-east division —the ten counties between the Moray Firth and the Forth—is 36·9° F., for the south-east 37·1°. For

Clackmannanshire, therefore, the most southerly county
of the north-east division, $37°$ may be taken as the correct

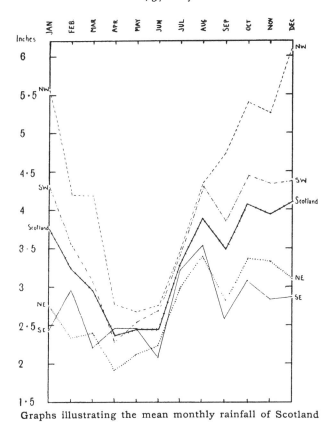

Graphs illustrating the mean monthly rainfall of Scotland

figure. The July temperature is similarly found to lie
between $56\cdot9°$ and $57\cdot9°$, and may be assumed to be $57\cdot4°$.

This gives an annual range of temperature for the county of 20·4°, being, as one would expect from an eastern county, somewhat above the mean range for the whole of Scotland.

The total rainfall for each of the four divisions and for the whole of Scotland and the distribution of this rainfall throughout the year are shown in the graph (page 35), which brings out clearly the greater rainfall of the west, especially the north-west, and the comparative dryness of the early summer months in Scotland. The amount of bright sunshine received has a close connection with the amount of rainfall, tending to vary inversely, especially so when, as is the case here, the summer months are the drier months. We find therefore that May and June are the brightest months of the year, receiving each nearly 200 hours of bright sunshine, except in the north-west where a greater degree of cloudiness obtains. If, as before, we attempt to get figures for Clackmannanshire as a mean between those for north-east and south-east Scotland, the total number of hours of bright sunshine appears to be 1490 per annum, the monthly minimum being 35½ hours in December and the maximum 185 hours in May. The total mean annual rainfall for the county would be 32·7 inches.

Within the limits of Clackmannanshire there are, of course, considerable local variations of climate; the broad distinction being between the higher, more exposed, colder and wetter Ochils and the lower, sheltered, warmer and drier plain. The difference in elevation between the portions north and south of the Ochil fault is sufficient

in itself to account for a difference of some 5° in temperature ; and, since the southern part is sheltered from
the north winds, the actual difference would be greater.
The southern slopes of the Ochils benefit equally with
the plain in being sheltered from the north and, moreover, present a surface more nearly normal to the rays
of the sun.   The mean annual rainfall at Alloa is 32·46
inches ; but nearer the hills the rainfall is heavier, being
39·97 at Alva House and 41·66 at Dollar.   On the hills
themselves the amount is considerably greater than this,
and, at times of heavy rainfall, the rivers rise rapidly and
come down in a huge spate, while the boggy morass called
Maddy Moss has been known to burst its barriers and
send down a jostling torrent of mud by the Burn of
Sorrow to the Devon below.   These Lammas floods
have often caused the Devon below Dollar to overflow
its banks, destroying any crops in the low fields adjacent.
Such a flood is noted in the *Old Statistical Account* :
" A very remarkable and uncommon flood happened in
September 1785, which carried away a prodigious quantity
of corn, broke down a stone bridge at the Rack Mill, in
Dollar, and occasioned other very extraordinary damage.
The river rose in four or five hours more than 13 feet
above its usual height at Tillicoultry Bridge." A later
and more disastrous flood occurred in August 1877.
That month had been one of particularly heavy rainfall,
so heavy, indeed, that it caused the total rainfall for that
year to be greater than previously recorded in many parts
of Scotland, including Dollar, where the total fall was
61·28 inches as against the average figure of 41·66.

Waterfall in Tillicoultry Glen

Throughout the night of the 27th August, a steady fall of rain had swollen the Ochil burns and, when morning came, a torrential downpour was descending from the gloomy and sombre sky. The dark hillside was streaked with the white of foaming, tumbling cataracts, which, pouring downwards into the turgid flood of the Devon, caused widespread inundation and devastation. The town of Tillicoultry suffered severely— to the extent, it is said, of some £8000.

8. Agriculture.

The 1911 Census returned the number of persons engaged in agriculture in Clackmannanshire as 490, one out of every 19 occupied males. Considerably less than half the land is now under cultivation, while one-third is mountain and heath land used for grazing. Oats and rotation grasses occupy more than two-thirds of the arable land, the remaining third being utilised for root crops, beans, wheat or barley. The soil south of the Ochil fault varies from a stiff clay to a sandy loam, the heavier soils being near the Forth, and the lighter, drier soils on the higher ground. In many places, however, the soil is peaty, and this accounts for the extensive cultivation of oats, since this crop, like potatoes, is more tolerant of extreme amounts of acid humus in the soil.

Before the close of the eighteenth century, the agriculture of the county was carried on in a wretched and wasteful manner. The holdings were so small as barely to furnish a subsistence to the family of the tenant, who, unable to carry out drainage operations on a suitable scale, found much of his cold, wet, and stiff clay soil not worth the trouble of working. Crude methods of cultivation, the absence of fences, the abundance of whins and weeds, miserable farm buildings and bad roads were

Harviestoun Castle

among the causes which rendered the farmer's existence poverty-stricken and precarious. The improvement when it came, proceeded apace. The size of the farms was increased as leases expired, in Clackmannan parish the number of holdings was reduced by 40 in 20 years; drainage and fencing were undertaken; in 1774, 300

acres of common grazing land belonging to the feuars
of Tillicoultry were taken by agreement for enclosure ;
thirlages and services were abolished ; a better rotation
of crops was introduced. Led by such men as Tait of
Harviestoun, the county rapidly acquired a reputation for
successful farming ; a Farmers' Society was established in
1784 ; three years later the first effective thrashing machine
in Scotland was erected at Kilbagie by George Meikle,
the son of its celebrated inventor, both of whom resided
at Alloa. In 1793 a plough and ploughman were sent
from the county to give an exhibition at the Royal Farm
at Windsor ; and, in the same year, one of the enclosures
of the Harviestoun estate was let at four guineas an acre
for grazing, whereas a few years before the best carse
lands of Clackmannan were rented at 43*s.* per annum.
This high price, however, did not continue.

The improvements so successfully and rapidly introduced have steadily continued ; but, since the middle of
the nineteenth century, there has been throughout the
country a contraction of the cultivated area year by year
and a tendency for arable land to revert to pasture. This
has been brought about mainly by the lowering of the
price of corn consequent upon the enormous increase in
imported grain. In seeking to discover to what extent
the area under cultivation in Clackmannanshire has con-
tracted, it is useless to take the actual acreage at different
times, because the county boundaries have altered, but
a comparison of the proportion of cultivated land to the
total area would be a sufficiently reliable index and such
is given below.

Area of Cultivated Land expressed as a Percentage of Total Area

1854	1881	1911
59	49	44

Area under various Crops or Grass expressed as a Percentage of Total Area under Cultivation

	1854	1881	1911
Wheat	7	5	2
Barley	7	7	1
Oats	16	21	19
Beans	6	4	2
Turnips	7	6	5
Rotation Grasses	21	22	24
Permanent Pasture	29	28	40

The Ochils provide excellent pasture for sheep, of which there were in 1911 15,220. There are some fine Leicesters on the best farms, but on the Ochils are kept black-faced ewes, crosses being bred from these with Leicester rams. The cattle number 3479, mainly cross-bred from Ayrshire cows and short-horned bulls, though many of the dairymen have bought Irish cows.

Only 3715 acres are now under woods and plantations, much less than formerly. Of the natural wood of the ancient forest, nothing remains but a very small portion within the grounds of Alloa House. The largest patches of woodland now existing are the "Forest" of Clack-mannan, at one time a royal domain, and Tullibody Woods.

9. Industries and Manufactures.

The chief industry of Clackmannanshire is the woollen and worsted manufacture, which now provides work for more than half the occupied females and for one-third of the occupied males. This manufacture, originating, doubtless, because of the large number of sheep on the Ochils, dates at least as far back as 1550, and was, for a long time, purely a domestic industry. The wool was hand-carded and spun by the wife, while the yarn was woven into cloth by the husband, the wife usually undertaking the purchase of the wool from the hill farmers and the sale of the cloth to the merchants. It was an advantage also to be near a stream, both for the purpose of cleaning the fleece and because the fulling of the cloth was done by the feet in the burn. Thus the industry gathered round the hill streams; Tillicoultry, Alva and Menstrie became important centres, Tillicoultry serges being famous throughout Scotland. These places were well situated to benefit by the later industrial developments, the hill streams provided water power, while supplies of coal were at hand. At first only blankets, serges and a coarse kind of plaiding were made; but, about the beginning of the nineteenth century, new enterprises were started; a successful tartan shawl manufacture began at Alva and Tillicoultry, and in the *New Statistical Account* the great increase of population in the latter place—from 1472 in 1831 to 3213 in 1841—is ascribed to the rapid increase in the shawl trade. In 1816 John Paton commenced the spinning

Tillicoultry from Devonside

of worsted yarns at Alloa, which has proved in recent years the most prosperous manufacture of all.   Alloa is now the great centre ; tweeds, it is true, are no longer manufactured there ; but all kinds of yarn are spun, the material used being chiefly colonial, Cheviot, or Saxony wools, as the wool from the black-faced sheep of the Ochils is too coarse.   The older centres, Tillicoultry, with nearly a dozen factories, and Alva, are still associated with the manufacture; and Messrs Paton have also a large yarn mill at Clackmannan.

Another industry is brewing and distilling, in which 831 persons are employed.   Brewing is an old-established industry ; the Sheriff's Account of 1359, states that the brewing women of Clackmannan were wont to pay an annual licence of fourpence.   Alloa is now the centre of this industry, and its ales have been famous since the first brewery was established in 1784.   To-day there are eight breweries with an aggregate annual output of over 100,000 barrels.   Spirits were formerly manufactured at Kilbagie and Kennetpans.   In the latter part of the eighteenth century, Kilbagie was the largest distillery in Scotland, with an annual output of 3000 tuns of whisky.   It was connected by canal and tramway with Kennetpans ; and the two distilleries are said to have paid in excise duties an amount greater than the whole land tax of Scotland at that time.   Indeed, in 1861, the excise duties on spirits and malt from Clackmannanshire contributed a sixty-eighth part of the revenue of the United Kingdom.   Changes in the law brought about the abandonment of these old distilleries ; the most important now existing are Carsebridge, established in 1799, one of the largest in the country, with a capacity for a weekly output of 60,000 gallons,—Glenochil near Menstrie, and Cambus.

The manufacture of iron provides work for 435 men. There is a foundry at Alloa, but the chief establishment was the Devon Iron Works near the Devon in the north

The docks, Alloa

of Alloa parish. These works, founded in 1792, were at first largely dependent upon local sources of supply, employing at one time 150 miners, but the seam of iron became exhausted. After experiencing fluctuating fortunes and frequent changes of ownership, the manufacture was abandoned some 40 years ago.

In addition to the industries common to all counties,

such as house building and furnishing, the preparation and sale of provisions, Clackmannanshire has bottle and glass works, shipbuilding yards, saw mills and rope walks at Alloa ; paper mills at Kilbagie ; brick and tile yards and pottery works at Alloa and Clackmannan.

10. Mines and Minerals.

The Clackmannan coalfield, which lies, of course, entirely to the south of the Ochil fault, is one of considerable value. It begins at the boundary with Fife, where the Millstone Grit dips gradually under the Upper Coal Measures. The Sauchie coalfield was the earliest to be worked, but its development was retarded by the lack of good roads to bring the coal to market. Towards the close of the eighteenth century the Devon Iron Works Company, then the tacksmen or lessees, found it hardly profitable to continue working the seams, since the only road to the port of Alloa was a semi-private one leading through the estates of the Mar family, by whom it was kept in repair and to whom had to be paid the Gate Mail, a toll of fourpence on every chalder of coal. The Clackmannan coal is now worked by the Alloa Coal Company, in whose deepest colliery 17 seams are found, aggregating 50 feet in thickness, five of these being under two feet. During the quinquennial period 1873–1877, the average annual output was 226,526 tons, since when it has almost doubled, the average annual output 1900–1904 being 421,448 tons. The output for 1911 was 414,746 tons.

The coal shipments from Alloa, where nearly half this output is shipped, are as follows:

1870	95,925 tons
1880	132,043 ,,
1890	300,921 ,,
1900	274,877 ,,
1911	214,826 ,,

The great increase in the shipments following 1885 was due to new docks and improved facilities for coaling; but the decline since 1890 seems to be the result of the development of the Fifeshire coalfields and the competition of the better situated ports below the Forth Bridge. In 1911, 1477 persons in Clackmannanshire were engaged in the coal industry; and the quantity of available coal from this field was estimated by the Royal Commission on Coal Supplies in 1905 at 443,800,366 tons.

No other minerals are at present worked in the county, though iron, silver, and copper were formerly mined, while lead, cobalt, arsenic, sulphur, and antimony occur in small quantities among the igneous rocks of the Ochils. The iron was mainly wrought in the vicinity of Tillicoultry; and, in 1793, the Devon Iron Works employed 64 miners and 10 women bearers in working the ironstone. A cave in the Mill Glen still marks an adit of an abandoned mine. In the Mill Glen, too, copper was wrought for a few years in the middle of the eighteenth century; a London company employing 50 miners till the veins, of which the thickest was about 18 inches, became exhausted. The little glen between the Middle and Wood hills behind Alva is

The Silver Glen, Alva

known as the Silver Glen; and here Sir John Erskine, in 1712, discovered his famous silver mine, which, for a time, was worked with considerable success, bringing to the owner nearly £50,000 worth of silver. The attention of the Government was called to this mine and Sir Isaac Newton, as Master of the Mint, assayed the silver. Sir John, being involved in the Jacobite rebellion, had been outlawed in 1715, but his sentence seems to have been remitted in consideration of the knowledge of the workings which he was able to impart to the Government, who sent down a German expert to investigate the mine. Though his report stated that the ore was exceedingly rich, the returns from the working rapidly dwindled to the vanishing point; and Sir John seems to have lost most of his gains by continuing the enterprise after the veins had become exhausted. A resumption of the mining 40 years later proved equally unprofitable, though Alva church gained a pair of communion cups made from the native silver.

From pits or quarries within the county were obtained in 1911, 4142 tons of gravel and sand valued at £305, 7584 tons of igneous rock worth £845, and 1520 tons of sandstone worth £323.

11. History.

The history of Clackmannanshire has been a quiet one; few great events have taken place within the county; armies have seldom encamped on its soil; it has not been the centre of great movements affecting the social or religious life of the country. Nor is the uneventful nature of the history of Clackmannan difficult to understand. The main current of Scottish history swept past the western boundary of Fife ; whatever military campaigns were undertaken, it was seldom necessary to attempt the seizure of the peninsula. One may search the session records of the Clackmannanshire parishes and discover scarcely any reflection of the general history of the country, even during its most eventful periods. Moreover, though Clackmannanshire is the southern door to Fife, the key to that door is Stirling. That town is famous as a place of great strategic import- ance ; its neighbourhood is thickly studded with the sites of battlefields. To the holder of Stirling, the possession of Fife was not a matter of urgent importance. The military history of Fife, since the incursions of the Northmen, is meagre in the extreme ; so also is that of Clackmannan, but the influential centres in the former county, from which light and learning spread over Scot- land, have no counterpart in the latter. In fact, the few incidents, worthy of mention in this brief history, are connected with men who passed through on their way to or from more eventful scenes.

It seems probable that, during the Roman occupation, some of the expeditions beyond the Antonine wall may have passed through Clackmannan. Roman coins have been discovered ; a double-edged straight iron sword, 31 inches long, was dug up near Harviestoun in 1796 ; cinerary urns have been found at Alva, Tillicoultry and elsewhere. In 1828, while an old road at Alloa was under repair, a supposed Roman burial-ground was discovered. Twenty cinerary urns of coarse pottery, rudely ornamented, were found, along with two stone coffins and a pair of gold penannular armlets.

In the early centuries of our era the men of this district belonged to the great Celtic tribe known to the Romans as Damnonii. Later, the county was part of the debatable land in the middle of Scotland, the district of Manann, inhabited by a mixed population of Britons, Angles and Picts, the scene of conflicts among the different peoples whose countries met here.

About the close of the seventh century, St Serf came to the Fife peninsula to convert the Picts. Many curious traditions linger round his name : of his long argument with the Devil at Dysart ; how he slew a dragon at Dunning ; how he raised from the dead two young men at Tillicoultry ; of the pathetic death at that place of his pet goat, or làmb ; how he rescued from the waters of the Forth an infant, who later became St Mungo. This much at least is true, that St Serf lived and that, from Culross as centre, he went out preaching and teaching, visiting, among other places, Tullibody, Tillicoultry and Alva. A well at Alva, a bridge over the Devon, and an island in Loch Leven, commemorate his name.

One of the greatest events of the ninth century was the union of the Picts and the Scots (844) under Kenneth MacAlpin. No precise and satisfactory account of the union is preserved. It is conjectured that Kenneth had a plausible claim to the Pictish crown, and taking advantage of an invasion of Pictland by the Northmen—perhaps in concert with them—he forced the Picts to submit. Tradition locates their defeat near Tullibody, where a stone on Baingle Brae was reputed to mark the battlefield. Kenneth's son Constantin had many troubles

Tullibody Church

with the Northmen. In 877 a swarm of them from Ireland penetrated the Firth of Clyde, crossed to the east coast, and defeated him at Dollar. He fled, but was brought to bay in the Fife parish of Forgan, where he fell with many of his men.

In 1559 French troops were on the coast of Fife in the service of the Queen Regent. Their commander, hearing that the English fleet sent to aid the Lords of the Congregation had arrived in the Forth, decided to retreat towards Stirling. To impede his march, Kirkaldy of Grange broke down the bridge of Tullibody over the Devon, then swollen by winter floods. The French, however, unroofed the church of Tullibody and repaired the broken bridge.

Much greater disasters to the county attended the passage of the royalist force of Montrose in 1645. He burnt Muckart and Dollar and also Castle Campbell, then one of the seats of Montrose's bitter enemy, the Marquis of Argyll. Alloa having been plundered by Montrose's lawless Irish auxiliaries, the main body spent the night in Tullibody Wood, while their leader and his chief officers were lavishly entertained at Alloa House by the Earl of Mar. Argyll, who was in pursuit, by way of reprisal burnt Menstrie House, the seat of the Earl of Stirling, and Airthrey Castle belonging to Sir John Graham of Braco, Montrose's uncle. He also threatened to return and burn Alloa House to teach the Earl of Mar to be more careful in his choice of guests. Though Montrose's victory at Kilsyth removed any immediate danger to Mar, yet, after Montrose's complete defeat at Philiphaugh by David Leslie, it was only the victor's direct intervention that saved the Earl's life.

12. Architecture.

The oldest castles now standing in Clackmannanshire go back to the Wars of Independence. Built at a time when the kingdom was in a disturbed state and when the great barons had neither the means nor the inclination to erect costly ornamental edifices, the early castles were especially designed for strength in defence. Their general plan of construction is similar, and shows the influence of the Norman ideas introduced into Scotland either by Anglo-Norman immigrants or by those who, during their raiding expeditions into the north of England, had seen Norman keeps. Examples of such imitations are Castle Campbell, Alloa Tower, and Clackmannan Tower. Generally speaking, these and similar buildings consisted of a square or oblong tower with walls of great thickness built entirely of stone. The walls of Alloa Tower are ten feet thick and a convincing proof of their ability to withstand attack by fire was given in 1800 when, a fire having accidentally broken out, only the ancient tower survived the conflagration, while the extensive buildings which had been added were burnt to the ground. These square towers were defended from the roof, which had a parapet supported on corbels. In some cases, as at Castle Campbell, small overhanging turrets—bartizans—projected from the angles of the battlements. In some cases the parapet had openings through which molten lead or other material might be cast on the heads of assailants. Such openings, styled machicolations, are, however, unusual in Clackmannanshire. The entrance to the tower was

usually made to the first floor level by a ladder or some
other easily-removable means. The ground floor was
used as a store-room and, like the other floors, was vaulted.
The first floor was the common hall and the second
contained the private rooms, while, if there were four
storeys, the top floor contained attics for the garrison,
entering from the battlements. These floors were connected with each other by means of a straight flight of
stairs in the thickness of the walls or by a newel stair,
i.e. a stone staircase winding upwards inside an upright
column in the angle of the walls. This was the usual
type of building and one so constructed might well defy
the most determined onslaught, more especially as the
towers occupied positions naturally strong, as the top of
a precipitous hill, and were surrounded by a courtyard
within a wall of enceinte. These towers, however,
provided but cramped and inconvenient accommodation.
As time went on, the need for defence became less urgent
while the desire for further accommodation became more
imperative. Wings were added, making the buildings
approximate to the plan of a mansion built round a
quadrangle. In the examples in Clackmannanshire, the
old quadrilateral tower is seen to be the main body of the
building, and the later additions, more extensive and, in
one case, higher than the original tower, are merely
adjuncts to it.

Probably the oldest castle in the county, as it is
certainly the finest in situation, is Castle Campbell. It
stands, one mile north of Dollar, on the crest of a rock
surrounded by deep wooded ravines on all sides except

the north, its only accessible side in former days. A
modern pathway up the glen to the castle was opened in
1865. The Castle passed, by marriage, into the hands
of the Campbell family about the middle of the fifteenth
century ; and the new owner obtained in 1489 an Act
of Parliament changing its old name of Castle Gloume.
The chief incidents in its history were the visit of John

Castle Campbell

Knox in 1556, the burning of the castle by Montrose in
1645, and its garrisoning by Cromwell's soldiers in 1652.
The little knoll on the green sward in front of the castle,
and between it and the edge of the cliff, is reputed to
be the place where the great Reformer dispensed the
sacrament of the Lord's Supper. After the Castle's
partial destruction by Montrose, it was allowed to fall

into disrepair; in 1805 the Argyll family sold the property to the Taits of Harviestoun, who sold it to Sir Andrew Orr in 1859. The original keep of Castle Campbell shows all the regular features mentioned above. It is a plain quadrilateral structure of four storeys, three of them vaulted; while the parapet supported on a corbel course has rounded bartizans with carved gargoyles. From the ground level, a straight flight of stairs in the thickness of the wall leads to the first floor, across which, in the opposite wall, another staircase leads to the upper floors. The horizontal position of the loopholes for firearms in the extension buildings shows them to be of later date. Other points of interest in the south wing are the fine entrance portico and the unusual feature of a corridor on the first floor connecting the two staircases and the apartments.

Alloa Tower, built about the commencement of the fourteenth century, was granted by David II in 1360 to Sir Robert Erskine in exchange for Strathgartney in Perthshire. The old newel staircase in the south-west angle still remains and the top floor with the battlements preserves its original features; but otherwise the tower, especially the interior, has been considerably altered. Wings were added later to the old keep and in time an extensive range of buildings arose. The Erskines, in whose hands the property remains to-day, were for more than a hundred years the guardians of the royal heirs to the throne of Scotland; at Alloa Queen Mary spent part of her childhood, and here also, in later years, were brought up her son James, and James's heir, the young

Prince Henry. The present seat of the family, Alloa
Park, was erected in 1838. The white freestone, of
which this handsome mansion in the Grecian style is built,
was obtained from quarries on the estate.

Clackmannan Tower, whether built, as tradition
asserts, by King Robert Bruce or not, was certainly a
royal domain till 1359, when David II granted it to a

Clackmannan Tower

relation, Sir Robert Bruce, in whose family it remained
till 1791. It is now the property of the Marquis of
Zetland. The oldest portion of the building is the rectangular tower on the north side. The south wing was
added about 1500 as the style of the fine fireplace on the
second floor helps to prove ; and it is a unique feature
that the new wing was carried to a height greater than

that of the original keep. Later additions of the seventeenth century include the walls round the forecourt with the moat and drawbridge in front, the belfry on the watch turret, and the Renaissance arch and entablature of the east doorway.

Besides providing additional accommodation by adding wings, the later proprietors sometimes erected buildings round the inside of the walls of enceinte. An example of this occurs at Old Sauchie Tower, where a house was added to the west wall in 1631. The keep itself, standing in the grounds of Schaw Park, was erected about the middle of the fifteenth century by Sir James Schaw, whose family arms and motto "I mein weill" may be seen in the tympanum above the entrance door in the western wall. The motto which can, perhaps, be made out in the tympana above the dormer windows to the left and right of this door, is "En bien faisant" on the left, continuing "je me contente" on the right. The property remained in the hands of this distinguished family for nearly four hundred years. In 1752 Sir John Schaw died leaving an only child as heiress. She married Charles, Lord Cathcart, whose grandson sold the property to the present owner, the Earl of Mansfield.

Little is now left of the once large mansion of Menstrie Castle, where Sir William Alexander, the first Earl of Stirling, was born. He was Secretary of State for Scotland, 1626–1640, one of the leaders of the enterprise for colonising Nova Scotia, and a poet of some merit. The only part of Menstrie Castle now fairly well preserved is the south front, where the entrance gateway is worthy of notice, exhibiting a somewhat curious example of that mixture of Scottish and Classic architecture which is characteristic of the seventeenth century.

The chief mansions in the county, besides Alloa Park and Schaw Park, are Kennet House, built in the early years of the nineteenth century, the seat of the Bruces of Kennet, who, in 1868, established their claim to the Barony of Balfour of Burleigh; Alva House; Harviestoun Castle, built in the Italian style in 1804, a new porch and tower being added in 1860 (see p. 40); Tillicoultry House, built 1806 ; Tullibody House.

The ecclesiastical architecture of the county is not of great interest. Tullibody church was built by David I in 1149, but Tullibody is no longer a separate parish and regular service has long been discontinued in the old building. The existing parish churches of Clackmannanshire are all modern buildings dating from the first half of the nineteenth century and are mostly Gothic in style.

13. Distribution of Population. Communications.

The courses followed by the roads and the positions of the urban centres of population are, in a sense, mutually dependent. Modern roads—improved highways representing older beaten tracks—have been constructed for the purpose of connecting the larger centres of population; but it is also true that a highway once constructed attracts inhabitants. Other things being equal, people prefer to live close to the highway, where they can most readily obtain the benefit of easy communication. In Clackmannanshire, generally speaking, the roads have influenced the positions of the towns rather than the towns the direction of the roads. From Stirling roads lead into the Fife peninsula along the narrow stretch of flat ground between the river Forth and the steep southern face of the Ochils, but, where this narrow passage opens out eastwards, the ridge of the Cleish Hills splits it in two. Hence there are two main roads from Stirling into Fife : one north of the Cleish hills through Kinross, and one south of the Cleish hills through Dunfermline. The towns of Clackmannanshire are found strung out along these two roads. Were there no other considerations to be taken into account, the towns might be expected at approximately equal intervals ; they would be small urban accretions, each with its church, school, inn, and shops serving the more essential needs of the immediate neighbourhood. A glance at the accompanying sketch map shows that the actual distribution of towns does fulfil such an expectation, but there are some further considerations to be noted. The place where the Forth ceases to wind is peculiarly suitable for the growth of a river port and here, in fact, grew up Alloa, a town which now holds more than half the burghal population of the county. In order to include Alloa, the southern or Dunfermline road leaves the direct line to bend southwards. The precise positions of the towns on the northern or Kinross road are obviously determined by the hillside burns. An abundant supply of pure water is always a desideratum, and was especially favourable to the woollen industry. A secondary system of roads connects the towns on the northern highway with those on the southern.

That the railway lines do not exactly correspond to the main roads is a consequence of the time and order

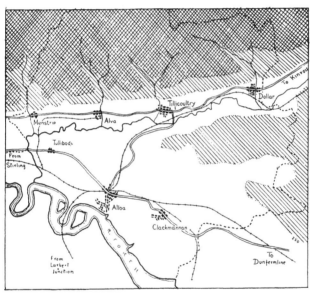

Sketch-map of Clackmannanshire, showing the influence of the physical features on the lines of communication

of their development. The railway corresponding to the easier southern road was the first to be constructed. This was the Stirling and Dunfermline railway opened to traffic in 1850; with, in 1851, a branch line from Alloa to Tillicoultry, at that time a prosperous town, whose rapid growth promised considerable traffic. When the more difficult construction of the railway corresponding to the northern road along the Devon valley was undertaken, the promoters, finding Tillicoultry already connected with Alloa, preferred to start their line from Tillicoultry instead of from Stirling. Thus the Stirling-to-Tillicoultry section of the north road has no railway corresponding ; though, in 1863, the gap was partly filled up by the short branch which, leaving the Dunfermline line at Cambus, runs through Menstrie to Alva. The Devon valley line was finally opened to traffic on the completion of the difficult section from Dollar to Rumbling Bridge in 1871. In 1875 this line was absorbed by the North British Railway. In 1883, the Caledonian Railway obtained access to Alloa by the bridge over the Forth. Since 1815 there has been regular communication in the summer months by steamer between Stirling, Alloa and Leith.

A table showing the distribution of population by parishes is appended :

Density of population per square mile, by parishes

Parish	Density in total area	Density in extra-burghal portion only
Alloa	1530	522
Alva	470	81
Clackmannan ...	197	197
Dollar	167	53
Tillicoultry ...	415	147

The parish of Alloa, which occupies rather more than a fifth part of the county, contains more than half (55 %) of the total population of Clackmannanshire. It is the

Alloa from the south

only parish whose population has increased during the last intercensal period. The burghal portion of the inhabitants of the county is almost exactly double the extra-burghal.

14. Administration.

In former times, the local government of the county was largely in the hands of the sheriff, who was both a judicial and an administrative officer. He possessed both criminal and civil jurisdiction. His administrative functions were partly military and partly financial : he had command over the militia and was responsible for its efficiency ; he had to collect and account for all feudal casualities, fines, and forfeitures. Later, though the office of sheriff was maintained, the real executive power passed, both in England and Scotland, into the hands of the resident county gentry. In Scotland, when acting in an administrative capacity, these were known as Commissioners of Supply, a body first established in 1667. As Commissioners of Supply, all were enrolled who had the necessary qualification, which varied from time to time but was always based on the possession of landed property. The whole structure of local government was reformed by the Acts of 1889 and 1894. The Local Government (Scotland) Act of 1889 established an elected County Council in every county. The chief duty of the Commissioners of Supply now is to appoint one-half of the members of the Standing Joint Committee, which controls police affairs and capital expenditure. All other administrative functions of the Commissioners were transferred to the County Council, which elects the other half of the Standing Joint Committee. The County Council also took over the powers of the County Road Trustees and became the Local Authority under the Public Health Acts. It now has the general oversight

County Buildings, Alloa

of the local government of the whole county with the exception of the burghs. For public health administration the counties are divided into districts, in each of which a District Committee is appointed, consisting of the county councillors for the district together with a representative from each parish council.

The Act of 1894 established parish councils whose

chief duty is the management of poor relief, though they have also the control of footpaths and the power to provide recreation grounds, baths, libraries, and cemeteries. The administration of primary education belongs to school boards elected *ad hoc*, that of secondary and technical education is under the County Education Committee. This Committee is elected by the County Council and the chairmen of the school boards.

The earliest sheriff's account for Clackmannanshire in the Exchequer Rolls of Scotland is Sir John Menteith's account for 1348–1359. It shows that the Crown then possessed considerable property within the county, for the receipts include the rents of several estates, crofts and orchards and a certain sum for the foggage of the Forest of Clackmannan. The disbursements include several royal gifts and also the fee to the sheriff as Constable of the Tower of Clackmannan. The office of sheriff was hereditary in the family of Menteith from the time of Sir John till 1631 ; in the family of Hope of Wester Granton, 1638–1666, and in that of Bruce of Clackmannan, 1666–1693. The last hereditary sheriff was the Earl of Dumfries, to whom, on the passing of the Heritable Jurisdictions Abolition Act, 1748, a sum of £2000 was paid as compensation for the loss of his office. The modern sheriff is now a judicial officer appointed by the Crown. Clackmannan shares a sheriff-principal with Stirling and Dumbarton, and a sheriff-substitute with Kinross. Nearly all the administrative functions formerly exercised by the sheriff are now vested in the County Council while his position as head of the territorial military forces has passed to the Lord Lieutenant. Clackmannan provides three companies (E, F, H) of the 7th Battalion of the Argyll and Sutherland Highlanders with a strength of seven officers and 270 men out of an establishment of 10 officers and 353 men.

There are four police burghs in the county : Alloa, Alva, Dollar, and Tillicoultry. Clackmannan was formerly a burgh of barony, but is now a "special district" for water, drainage, scavenging and lighting. For purposes of ecclesiastical administration, the parishes are in the presbyteries of Stirling and Dunblane within the synod of Stirling and Perth. The county is united with Kinross-shire for parliamentary representation.

15. Roll of Honour.

The Erskines of Alloa, the Menteiths, the Bruces of Clackmannan and their related house of Kennet, the Taits of Harviestoun, the Abercrombies of Brucefield and Tullibody, the Johnstones of Alva—all had a long and honourable connection with the county, a connection which, in some cases, still remains unbroken. It is possible here to record the achievements only of the most renowned scions of these houses.

The connection of the Erskine family with the Earldom of Mar is somewhat obscure. The direct line of the old Earls of Mar failed in 1374 ; and in 1391 Sir Thomas Erskine claimed the earldom in right of his wife. Their descendant John, the sixth Lord

Erskine, received the title of Earl of Mar in 1565. Perhaps the most interesting member of this family is the Jacobite Earl, born at Alloa House in 1675. He was a Privy Councillor under William III; in Anne's

The Jacobite Earl of Mar

reign he changed sides more than once; he was one of the chief promoters of the Union of 1707, which he soon professed to regret; for years before Anne's death he was

in communication with the Pretender, but he was en-
thusiastically ready to welcome the Hanoverian King.
Mar's extraordinary versatility, even in an age when
changing sides was no rarity, gained him the nickname of
"Bobbing John." George I deprived him of his post as
Secretary of State and of the hereditary guardianship of
Stirling Castle. Under the Pretender's orders, Mar,
August 1715, left London for Scotland to stir up war.
On the 6th September the Jacobite standard was raised
at Braemar. Perth was seized on the 16th, but the
indecisive battle of Sheriffmuir, 13th November, proved
fatal to the Jacobites. James landed at Peterhead on
the 22nd December, but the absence of French support
rendered the rising abortive ; and, early in February, the
Prince, Mar and other leaders escaped to France. Mar
never returned and died at Aix-la-Chapelle in 1732.
The earldom was restored to his grandson in 1824. The
present title of the head of the family is Earl of Mar and
Kellie.

Of the Abercromby family, the most famous member
was Sir Ralph, who was born at Menstrie in 1734 and
had a distinguished military career in Flanders, the
West Indies, and Egypt. When in command in the
Mediterranean, he received orders to expel the French
from Egypt, and, landing at Aboukir Bay on the 8th
March, 1801, he was completely successful in resisting
a violent attack by the French general, Menou. Un-
happily, Sir Ralph was wounded in the thigh and died
seven days later. Sir Ralph's third son James was elected
Speaker of the House of Commons in 1835 and, on

Sir Ralph Abercromby

relinquishing office in 1839, was created Baron Dunfermline. Hence the claim, which was inscribed on an old village flag, that "Tullibody has served its country with valour in the field and with firmness in the Senate."

Another famous soldier was the first Lord Colville of Culross, to whom the estate of Tillicoultry belonged. He served under Henry of Navarre against the Catholic League. Raised to the peerage by James VI in 1609, he spent his latter years at Tillicoultry, where he died in 1620

Of Churchmen, first mention must be made of the devout and kindly Vicar of Dollar, Thomas Forret, who made it a practice, unusual for incumbents in the sixteenth century, to preach every Sunday to his parishioners. For this he was denounced to the Bishop of Dunkeld as a heretic. Forret's friend, Thomas Locklaw, the last Romish priest to officiate in the church of Tullibody, also held unorthodox views; and, in particular, renouncing celibacy, was married during Lent. Forret and four others were charged with being present at the marriage and then eating flesh, besides being chief heretics and teachers of heresy. Locklaw escaped to England; but Forret was burned in the presence of the King on the Castle Hill at Edinburgh, in the spring of 1539.

Another famous ecclesiastic was Archibald Campbell Tait, the son of Crawford Tait of Harviestoun. Archibald Tait, after being Headmaster of Rugby, Dean of Carlisle, and Bishop of London, was enthroned Archbishop of Canterbury in 1869. His life work, therefore, was done beyond the borders of the county, a contrast to the case of his father, Crawford Tait, who devoted his life to the agricultural improvement of Clackmannanshire, and, indeed, is said to have almost ruined himself with his agricultural experiments.

To Art, to Science, and to Commerce, the county has contributed many eminent men; but we must be content to select one representative from each sphere. David Allan (1744–1796), born at Alloa, was enabled through the generosity of the county families to study art in Rome, where his picture, "The Origin of Painting or the Corinthian Maid drawing the shadow of her lover," won the gold medal of St Luke's. After a few years' portrait-painting in London, he settled in Edinburgh and was appointed, in 1786, Master of the Edinburgh Academy of Arts. He now devoted his talents to depicting the manners, customs, and scenes of his own country. The "Scotch Wedding" and the "Highland Dame" are perhaps the two best examples of his art.

Robert Dick (1811–1866) was born at Tullibody, and in 1830 set up as a baker in Thurso. He devoted his leisure time to the study of nature, and gradually acquired a thorough knowledge of the botany and geology of the north-eastern corner of Scotland. No works of his were ever published; but his authority was widely recognised and Hugh Miller warmly acknowledged his invaluable aid.

John M'Nabb (1732–1802), born near Dollar, started life as a herd-boy. He migrated to London, and during a long life spent in sea-faring and ship-owning amassed a large fortune. On his death, he left £56,000 to endow

"a charity or school for the poor of the parish of Dollar."
Some difficulty was encountered in interpreting his
wishes ; but eventually in 1819 the Dollar Institution
was erected. It has become one of the great schools of
Scotland, and numbers among its distinguished alumni
Sir James Dewar, Sir David Gill, and the Swettenhams
of Malay fame.

16. THE CHIEF TOWNS AND VILLAGES OF CLACKMANNANSHIRE

(The figures in brackets after each name give the population in
    1911, and those at the end of each section are references
    to pages in the text.)

Alloa (11,893), a police burgh and the largest town in the
county, is the administrative and commercial centre. At Alloa
the neap tide rises some 15 feet, the spring tide about 23 feet, the
port having the additional advantage of double, or, as they
are locally called, "leaky" tides. This return of the flood tide,
after the water has sunk downwards for a foot or more, appears
to be due to the constriction of the river channel at Queensferry
hindering the free egress of the ebbing water. Tradition, scorning
so mundane a reason, asserts that the extra tide has persistently
recurred since the day when it was first sent to the help of
St Mungo, whose vessel had run aground on its way to Stirling.
Alloa was made an independent port in 1840 and its district now
includes the smaller ports or creeks of Charlestown, Clackmannan
Pow, Inverkeithing, Kennetpans, Kincardine, St David's, and
Stirling. The Board of Trade returns for the district show that
in 1911 the value of the total exports—almost exclusively coal—
to foreign countries was £81,718, and the value of the total
imports from foreign countries was £170,440, of which wood and
timber accounted for £128,450, stones and slates for £41,202.
The last item first appeared in the returns for 1909 and represents

stone imported at Limekilns for the Rosyth Naval Base. In recent years Alloa has benefited greatly by the munificence of the Paton family. (pp. 3, 8, 9, 21, 25, 37, 41, 45, 46, 47, 48, 52, 54, 58, 62, 64, 69, 74.)

Alva (4332), a police burgh three miles north of Alloa, is engaged in the woollen industry. The name, formerly spelled Alveth, is derived from a Gaelic word *ailbheach*, "rocky." (pp. 8, 23, 29, 43, 45, 50, 51, 52, 64, 69.)

Cambus, a village at the confluence of the Devon and the Forth, noted for its distillery. The name is Gaelic, "the bend of the water." (pp. 26, 46, 64.)

Clackmannan (1385), a small town on the Black Devon river, at the mouth of which river is the creek of Clackmannan Pow. The town, now dwindling in size and importance, still claims to be considered the county town. At the top of the old cross are carved the arms of the Bruce family, whose chief seat for many generations was Clackmannan. (pp. 3, 8, 29, 45, 47, 59, 69.)

Coalsnaughton, a collier village about a mile to the south of Tillicoultry.

Devonside, a village, on the Devon, between Tillicoultry and Coalsnaughton. It is united with the latter as a special district (pop. 1353) for water and lighting, but Devonside with its brick and tile works dates only from the middle of the nineteenth century, whereas Coalsnaughton or Coalsnachton has quite a long history.

Dollar (1497), the smallest of the four police burghs, is situated close to the point where the Devon enters the county. Its beautiful and healthy situation and the educational advantages afforded by the Dollar Institution combine to make Dollar a residential town rather than an industrial centre. It has extensive bleachfields. The famous engineer, James Watt, reported that the Devon might be made navigable as far as Dollar at an estimated cost of £2000; but nothing was ever done to carry out the scheme. (pp. 8, 22, 25, 37, 53, 54, 56, 64, 69, 74, 75.)

Forest Mill, a small hamlet on the north margin of the Forest at a point where the Clackmannan and Kinross road crosses the Black Devon. Michael Bruce, the young author of *Lochleven*, was schoolmaster here in 1766. (p. 136.)

Kennet, a mining village, one mile east of Clackmannan.

Dollar Institution

To the west of the village is Kennet House, the family seat of the Bruces of Kennet. (p. 61.)

Kennetpans, a small village on the shore of Clackmannan parish, with a small harbour. (pp. 25, 45.)

Kilbagie, a small village in Clackmannan parish, was formerly noted for its distillery; but at the present time its only industry is paper-making. (pp. 45, 47.)

Menstrie (667), a village in Alva parish, on the Stirling-to-Kinross road, is engaged in the woollen manufacture. Like the other villages and towns along the foot of the Ochils, Menstrie has a very beautiful situation. (pp. 8, 23, 29, 43, 60, 64, 71.)

Sauchie, a village on the Alloa-to-Tillicoultry road. There are also the villages of Old Sauchie near the Devon, and New Sauchie on the outskirts of Alloa. The population of the special scavenging and lighting district of Sauchie is 2844, and the people are mostly workers in the coal mines. (pp. 47, 60.)

Tillicoultry (3105), a police burgh two miles east of Alva and, like that town, engaged in the woollen industry. There is now little vestige of the stone circle which formerly stood on the Cuninghar, a rising ground behind the town. (pp. 8, 23, 29, 39, 41, 43, 45, 48, 51, 52, 64, 69, 73.)

Tullibody (838), in the east of Alloa parish, is one of the oldest villages in the county. The old church is the mausoleum of the Abercrombies and contains many tributes to the various members of that distinguished family. (pp. 8, 52, 53, 54, 61, 73, 74.)

KINROSS-SHIRE

1. County and Shire[1]. Origin and Administration of Kinross=shire.

The origin of the shire of Kinross is lost in antiquity. Sir Robert Sibbald, writing at the close of the seventeenth century, says that it was made a distinct shire from Fife about the year 1426 and contained then only the parishes of Kinross, Orwell, and Portmoak. In 1426, Kinross was separately represented in parliament, but the shire was distinct far earlier, being mentioned in the charters of David II. The very name of the shire indicates that this part of the peninsula was regarded as, to some extent, separate from the rest, for Kinross means the head or mountainous part of the peninsula as contrasted with Culross, the back or lowest part. In 1685, the hereditary sheriff of this small shire was the celebrated architect Sir William Bruce of Kinross, who was able, by the favour of the King, to obtain an Act adding the parishes of Cleish and Tullibole. Subsequently, further changes took place ; and in 1891, when the Boundary Commissioners came to simplify the county areas, they found only two parishes, Orwell and Kinross, completely contained within the shire, five other parishes contributing to make up the

[1] See p. 1.

whole. Kinross-shire now contains five parishes: Kinross, Orwell, Portmoak, Cleish, and Fossoway, the last-named including the older Tullibole.

Like other counties[1], Kinross has a County Council, but, as the shire is too small for division, the whole Council sits as a District Committee with the representatives of the parishes, the latter withdrawing when the business in hand is not concerned with highways or public health. The county is divided into 20 electoral divisions, each returning one member to the County Council. The only police burgh in the county is Kinross, but Milnathort is a "special district" for water, drainage, scavenging and lighting.

Kinross shares a sheriff-principal with Fife, and a sheriff-substitute with Clackmannan. The official head of the county is the Lord Lieutenant, who is also President of the Territorial Association established by the 1907 Act. The Territorial force for the county of Kinross is G Company of the 7th Battalion of the Argyll and Sutherland Highlanders with a strength, in 1912, of one officer and 111 men out of an establishment of three officers and 117 men. For the purposes of parliamentary representation the county is united with Clackmannan; for ecclesiastical purposes it is within the presbytery of Kinross and the synod of Fife.

[1] See p. 66.

2. Position and General Characteristics.

The county of Kinross forms part of a peninsula shut off from the rest of Scotland by the lofty barrier of the Ochils, and consists essentially of a lake basin within a rim of hills. One might expect that the most striking feature about such a position would be its isolation, more particularly since Kinross is one of the few Scottish counties not possessing coast-line. But the surrounding hills are notched in more than one place, and through these notches pass the roads which bring the county into touch with the outer world. Were these roads merely the paths by which wayfarers might enter or leave a lake basin secluded amidst the hills, they would have little effect in neutralizing the natural isolation of the position, but Kinross happens to lie athwart the Great North Road, which, coming from the south through Edinburgh to Queensferry, crosses the county to reach Perth and the north. Another great highway from Stirling and the west enters Kinross by the Devon valley and crosses the plain to pass by the Eden Valley to Cupar, St Andrews or, by the Tay ferries, Dundee. The intersection of these two great highways has brought to the county the advantages which always attend a position where traffic is focussed. Such advantages are by no means to be estimated merely by the number of persons who find employment in handling the traffic and in providing for the wants of travellers, though it may be noted in this connection that the Census Report of 1911 returns

Kinross from Auld Kirk Tower

132 men as engaged in the railway service, a not insignificant number since Kinross contains a smaller population than any other Scottish county. The real advantage of the cross-road position is indirect and difficult to evaluate; it lies in the fact that the district has better facilities for the disposal of its produce, more rapid and more efficient means of transport and communication with accessible markets, than the resources of the county would, in themselves alone, justify. It lies, also, in the stimulus imparted to the general life of the community by a constant intercourse with outsiders.

With the exception of coal mining in the extreme south, the interests of Kinross are almost entirely agricultural. Two-thirds of the surface of the county is under cultivation, rotation grasses and oats being most extensively grown. A peculiar feature about the agricultural holdings of Kinross is that so many of them are occupied by owners. Stock raising is the most important industry; there are almost as many cattle as people in the plain of Kinross; while, on the surrounding hills, over 34,000 sheep find excellent pasturage within the county boundaries.

3. Size, Shape, Boundaries and Surface Features.

The total area of the county is 55,849 acres or, excluding water, 52,410. Excepting Clackmannanshire, Kinross is the smallest county in Britain. The county is roughly oval in shape; measuring through the town of Kinross, the greatest length from east to west is 13 miles, from north to south nearly nine. It is bounded on the north and west by Perthshire, on the east and south by Fife, while in the south-west, an extension to the borders of Clackmannanshire breaks the regularity of the oval shape. Occupying the lowest portion of this oval basin is Loch Leven, famous for its trout, and, from its historical associations, the most interesting of all Scottish lakes. In shape resembling somewhat a tilted harp, this lake has a surface area of 3406 acres and a periphery of over ten miles. Its surface is 350 feet above sea-level and the volume of the lake, when the water is at full height, has been estimated at 600,000,000 cubic feet. Surrounding the lake lies the flat plain of Kinross, most extensive to the west, and across this plain flow numerous small streams, whose sources are to be found amongst the hills engirdling the lake basin. The hills approach the lake most closely in the south, where, about a mile from the shore, the steep face of Benarty Hill rises to a flat top more than 1000 feet above sea-level. The county boundary runs for some distance along the crest of the hill, here, as often elsewhere, coinciding with the water-parting—Kinross-shire being essentially the area draining into Loch Leven. Benarty resembles most of the hills in the Fife peninsula in having an abrupt descent on its western side, where a steep wooded slope overlooks the Great North Road and railway, which here cross the water-parting at a saddle only 420 feet above sea-level. West of this gap, the land rises to the long east-and-west ridge of the Cleish Hills, whose highest point, Dumglow, is 1241 feet; here again the boundary generally follows the water-parting.   Over the irregular surface west of the Cleish Hills, the boundary continues generally westwards to the Clackmannanshire border, where it turns north to the river Devon, up which it passes to the con-

Benarty from the north, showing the steep western slope

fluence of the little Glendey burn.   This burn runs along the southern exit of a transverse pass in the Ochils, the pass utilized by the road from Dunning to Muckart. Up this pass the county boundary runs towards the culminating point (1007 feet), approaching which, it strikes eastwards along the crest of the main ridge of the

Ochils. Here is Innerdouny Hill (1521 feet), the highest point in Kinross. East of this, the boundary follows a somewhat irregular course over generally falling ground until, at a point where the three counties of Fife, Perth and Kinross meet, it drops into the valley leading to Glen Farg and Perth. This valley is the exit used by the Great North Road and the railway which entered the county at the gap between the Cleish and Benarty Hills.

Glen Farg

The boundary line, having dropped into this valley, turns south-east towards the lofty Lomond Hills, following the course of the little Carmore burn, one of the head streams of the river Eden. In the gap between the Lomonds and the north-east extension of the Ochils, the water-parting, dividing the Eden basin from that of Loch Leven, is of very low elevation, being, indeed, less than 50 feet above the level of the lake. Having crossed this divide, the

boundary line winds among the hills so as to include
the Bishop Hill (1492 feet) wholly in Kinross but to
exclude the twin peaks of the Lomonds. It then descends
to Auchmuirbridge, beneath whose arches flows the river
Leven carrying the surplus waters of Loch Leven to the
sea. From Auchmuirbridge to Benarty the boundary line
was readjusted by the Commissioners in 1891. It now
follows the New Cut, i.e. the present channel of the
river Leven, for two miles, after which it circuitously
approaches the crest of Benarty. As we trace the course
of the county boundary, it is impressed upon us that
the oval plain of Kinross is upon all sides environed
with hills—the Ochils in the north-west, the Lomonds in
the east, Benarty and the Cleish Hills in the south—but
that, between these hills, low saddles or deep-cut passes
provide easy means of ingress and egress.

The plain of Kinross is more extensive to the west
of Loch Leven, and it is, therefore, across the western
plain that the largest streams of the county flow.
Numerous burns from the Ochils are gathered up by
the North Queich, which rises in Innerdouny Hill and,
after a course of some six miles, enters Loch Leven north
of Kinross. Immediately south of the town, the lake is
entered by the South Queich, whose course is longer than
that of the North Queich but whose volume is not increased by so many tributary streams. Skirting the
northern foot of the Cleish Hills is the Pow Burn, an
affluent of the Gairney Water, which drains the southern
portion of the county.

4. Geology[1].

The disposition of the hills in a rim round the basin of Loch Leven affords the clue to the character of the rocks of Kinross. The rocks which underlie the plain surrounding the lake are sandstones, generally of a dark brick-red colour but with intercalations of white, yellow, and other tints, belonging to the Upper Old Red Sandstone group. There are few good exposures of the rock on the plain itself; but, where it slopes upward to the hills, the sandstones may be studied in the channels of the brooks or even exposed in the face of the Bishop Hill or Benarty. Here false bedding may be noticed, while the sandstones, besides being studded with flat pellets of reddish clay, often contain fragmentary fish remains. Sandstone is comparatively easily disintegrated; hence any district where sandstone underlies the soil has usually been worn down, by the long-continued action of denuding agents, to a level lower than the circumjacent country. The average elevation of the plain of Kinross is not much more than 400 feet while the hills around never fail to reach beyond 1000 feet above sea-level, being built or more resistant rock, mainly the non-sedimentary rocks known as igneous or eruptive. Igneous rocks may be divided into two groups: volcanic and plutonic. Volcanic rocks, such as lavas, have been ejected through some orifice or fissure in the earth's crust and have cooled comparatively rapidly at the earth's surface. Plutonic rocks,

[1] See p. 11.

those which cooled slowly at a depth below the surface, are crystalline to a higher degree than volcanic rocks, and have only become exposed by the removal through denudation of the overlying material. Plutonic rocks are subdivided according to the form taken by the cooling mass ; the form being determined by the shape of the cavity into which the molten material was injected. Where the rock simply forced its way up some line of weakness, it cools either in the shape of a plug, and is called a "boss" ; or, if the line extends horizontally, it cools in the shape of a wall, and is called a "dyke." But sometimes the molten material penetrates, more or less horizontally, between the layers of stratified rock, spreading out in a sheet, when it is called a "sill." Examples of all these forms may be found in Kinross. Benarty is a great sill of dolerite (or greenstone), protecting the underlying sandstone. The Bishop Hill is part of a similar, but much more extensive sill, which, resting on and penetrating rocks of the lower Carboniferous series, forms the roughly circular mass of the Lomonds. The Cleish Hills consist of smaller sills considerably reduced in size, and so detached from each other, by denudation. Cowden Hill (933 feet) in this district is, however, a volcanic neck, that is, an orifice or funnel, drilled in the earth's crust by the volcanic explosion, up which the lava wells towards the crater. Subsequently the crater becomes choked with a varied assortment of fragmentary material, usually penetrated by dykes or veins of igneous origin. There are few of these volcanic necks in Kinross. The neck at Cowden Hill is filled with a greenish volcanic agglomerate and contains a plug of basalt. The Ochils, though composed mainly of igneous rock, are different in character from the Bishop, Benarty, and Cleish Hills. They are built up of thick sheets of agglomerates and lavas, the latter being the molten material ejected to the surface during the volcanic period contemporaneous with the deposition of the Lower Old Red Sandstone, while the former consists of fragmentary material due mainly to volcanic explosions. Penetrating this main mass of andesitic lavas and agglomerates are, in some places, found later intrusions of igneous magma, mostly dykes. One of notable length, starting to the south of Dochrie Hill in Kinross, runs along the southern slope of the Ochils in a north-easterly direction for a distance of 30 miles or more.

In Kinross the geological events of the Ice Age have considerably modified the topography. The great mass of the ice-sheet, grinding slowly eastward or south-eastward over the land, removed the most prominent irregularities, and smoothed and polished the rocks over which it passed. Thus to-day the hills retain a characteristically undulating form, and here and there scratches on the rock surface indicate the direction in which the ice moved. The low ground, when the ice melted away, was left clothed with a thick blanket of decomposed material, which had been swept downwards from the hills by the ice. This material, generally a clayey matrix full of fragmentary material, is known as Boulder Clay or Till, and is distributed irregularly over the whole of the Kinross plain, stretching also up the valleys of the Ochils, where it can be seen terraced and trenched by the descending streams. Over the plain itself the Boulder Clay is often found in smooth oval-shaped ridges known as drumlins. The ice also brought down great boulders, carrying them onward until, with the melting of the ice, they were dropped on rocks of quite a different kind. Many such

A Farmstead among the Ochils

boulders may be noticed in Kinross, especially on the Cleish Hills, where their character shows them to be gneisses or schistose rocks from the Highlands. Again, streams issuing from the glacier laid down fan-like deposits of sand and gravel. As the glacier retreated, these deposits followed, so that the deposits of successive periods formed in time a continuous gravel ridge, known as an "esker." One to the south-west of the town of Kinross, known locally as Drungie Knowes, can be traced for about two miles. Besides the eskers, irregular mounds or short ridges of water-deposited gravel or sand, called "kames," occur frequently in the Kinross plain, notably near Coldrain. Both the eskers and the kames overlie the Boulder Clay, but are less widely distributed than it.

The deposition of these glacial deposits formed numerous basins on the irregular surface and left the old lines of drainage partially obliterated. The basins filled with water and became lakes. The smaller ones—indeed, almost all—have since become converted into peat mosses, but Loch Leven remains as a conspicuous example. This loch, though still the largest lake of the Scottish lowlands, was once much larger. The flat terraces of alluvium, over which its waters once spread, rise to a height of at least 60 feet above its present level, and stretch from Milnathort in the north-west to Auchmuir in the south-east. This would give the lake an area of at least 5000 acres, about half as large again as its present size. The contraction in size is due partly to natural and partly to artificial causes. Besides the natural silting up of the lake, the overflow of the water, escaping by the river Leven through a narrow defile at Auchmuirbridge, has at this place so cut down its channel as to reduce considerably the height of the loch. Secondly, between 1826 and 1836, a new channel, some three miles in length, was cut at a lowel level from the loch to Auchmuirbridge. This, "the New Cut," permanently reduced the level of Loch Leven by 4½ feet. Power to cause a further reduction in level has been given to the mill owners along the river, who also benefit by the prevention of the disastrous floods which, previous to these operations, were frequent in winter time in the lower Leven valley. The depth of Loch Leven is, in two pits, greater than 60 feet, but this is more than twice

The New Cut, looking east

the average depth. It contains seven islands, whose sandy structure shows them to be the tops of kames. St Serf's Island is the largest, containing some 80 acres; while the only other island of any size is Castle Island, containing eight acres and situated off the town of Kinross.

5. Natural History.

Two-thirds of the county is under cultivation, with oats as the chief crop. Where the July temperature is below 56° F. and the rainfall exceeds 34 inches per annum, wheat ceases to be a regular crop; and as the July temperature of this shire is but slightly above this minimum and the rainfall is certainly greater than 34 inches, practically no wheat is sown. Cultivation in favourable localities on the southern slopes of the Ochils may be carried up to the 1000-feet contour line; but more generally the cultivated area passes about the 800-feet level into the zone of hill pasture. The lower slopes of the hill pasture are clothed with mixed grasses and, on the drier knolls, with whin. The upper slopes have *Nardus stricta* dominant, *Molinia varia* abundant, and on the drier knolls the whin is often replaced by heather. Where the rock is not covered with boulder clay and the soil is formed from the weathering of the black basaltic crags, the vegetation is richer; and Dr Smith mentions that a list of 40 species of plants was made in August 1900 on the crags of Dumglow (Cleish Hills) at a height of 1000 feet with a south-west exposure. This list includes *Festuca ovina, Geranium sanguineum, Trifolium pratense, Anthyllis vulneraria, Rosa spinosissima, Pimpinella saxifraga, Scabiosa succisa* and *Veronica officinalis*.

The woods of the county are coniferous (chiefly Scots pine) at Knock Wood, Tilliery Plantation and Leven-mouth Plantation; birch on the hillside north of Scotlandwell; oak on the west foot of Benarty; and mixed at Blairadam and in the grounds of Kinross House. Kinross House has some very fine beech trees.

The rabbit, the hedgehog, the vole, and the brown rat are abundant in the county; much less common are the hare and the fox, while the badger and the otter tend

Birch woods overlooking Scotlandwell

to become exterminated. The pole cat, formerly found here, has vanished; and, indeed, is nearly, if not quite, extinct in Scotland. The shire is rich in bird life; among the rarer species *Alcedo ispida* (the kingfisher) and *Loxia curvirostra* (the crossbill) may be mentioned; while the islets of Loch Leven are veritable sanctuaries for wild fowl. Formerly there were people who, engaging boats at Kinross ostensibly for fishing, would visit St Serf's Inch for the purpose of robbing the nests; but this was stopped by forbidding the boats to land except on a small fenced-off portion of that island. Some of the birds which at one time used to breed on the islands are no longer found. Sir Robert Sibbald, at the close of the seventeenth century, mentions a small island near St Serf's Inch called the Bittern's Bower; but the bittern is not known now to breed in any part of Scotland. It may be of interest to give the other birds in Sir Robert's list:—the common heron, snipe, teal, water-rail, kingfisher, coot, swan, sundry gulls, wild geese and wild ducks. Loch Leven is also the home of *Anthicus Scoticus*, one of the seven species of beetles peculiar to Scotland.

6. Climate[1].

Kinross and Clackmannan, adjacent to each other and both so small in area, have identical climates except for the slight differences caused by the topographical features. It is the influence of the hills surrounding the shire of Kinross that produces the slight peculiarities of the climate. A range of hills obstructs the free passage of the wind. Above the hill-tops the air, of course, blows freely onward, and causes an up-draught of that lower air checked by the ridge. On the windward side of the hills, therefore, the air is obliquely ascending. As it

[1] See p. 34.

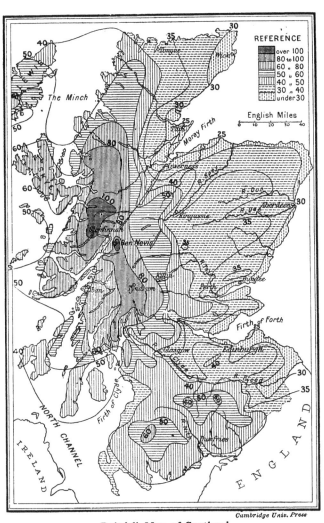

Rainfall Map of Scotland

(*By Andrew Watt, M.A.*)

ascends, the pressure of the overlying atmosphere is reduced; and the air expanding becomes less dense and, therefore, cools. The cooling of the air causes a certain amount of precipitation, and hence mountainous and hilly country always tends to experience a heavier rainfall than the adjacent piedmont plains. This heavy rainfall descends upon the windward slopes, as may be clearly seen in the rainfall map of Scotland, where the greatest rainfall is shown on the western slopes of the mountainous country, while that part of the Grampians east of the Spey and Tay valleys has, comparatively, a slight rainfall, proving that altitude is not so important a factor as exposure. In the case of Kinross, the disposition of the higher ground itself affects the direction of the prevalent winds and, in the absence of official statistics—for the Meteorological Office has no station in the county—we can only state generally that the west of the county gets more rain than the east—Mr Watt gives Blairingone on the Clackmannanshire border as having a mean annual rainfall of 39·80 inches, while Loch Leven Sluice has 36·59—and that the high ground gets more than the low. As a result of the combination of these two conditions, the Ochil portion in the north-west, the only area within the county where the mean annual fall is known to exceed 40 inches, has the heaviest rainfall, while the amount becomes less as one crosses the county to the south-east border, where, on the lower ground, it is as little as 35 inches per annum.

Besides influencing the distribution and amount of the rainfall, the high ground also affects the temperature.

The parish of Portmoak, for example, has a somewhat mild spring since the lofty wall of the Bishop Hill shelters it from the cold east winds, which often blow over Fife during that part of the year. Though the hills thus shelter the land on the lee side, they have in another way the effect of lowering the temperature. There is a decrease in temperature of 1° F. for, roughly, every 300 feet of elevation; and hence it is always colder or cooler on the hills than in the plains. In winter time winds from the north-west, reaching the plain of Kinross after passing over the often snowclad Ochils, are bitterly cold. When the figures are adjusted to sea-level, the average January temperature for the county is below 38°, that for July is $58\frac{1}{2}°$.

The presence of a large sheet of water and, before the extensive drainage operations of the early nineteenth century, the bogs and the damp undrained soil of the low-lying plain gave rise to a humid and somewhat unhealthy climate. Ebenezer Erskine, who lost four of his children during his ministry at Portmoak, speaks of the situation of the manse as "unwholesome," and it is a common remark in the *New Statistical Account* that ague, rheumatism, and pulmonary complaints were widely prevalent in the parishes of the county before the land was improved by drainage. The humidity of the atmosphere is still comparatively high; but the sufferers from these ailments are not now more numerous in the shire than in equal areas elsewhere in Britain.

7. Agriculture.

Agriculture and stock raising are the chief occupations of the people of the county. One out of every three occupied men finds work on the farms, the total number so engaged in 1911 being 756 men and 84 women. There are 298 holdings of an average size of 116·4 acres, and of these quite an unusually large proportion is occupied by the owners, as the following table shows:

Proportion of Cultivated Land occupied by Owners, 1911

Divisions of Scotland	Number of holdings occupied by owners, expressed as a percentage of total number of holdings	Acreage of land occupied by owners expressed as a percentage of total acreage
S.W.	10·79	11·38
N.W.	4·63	12·26
S.E.	12·68	12·45
N.E. (including Kinross)	7·80	11·64
Kinross	20·81	29·79

Corresponding to the 29·79 of Kinross, the percentage for all Scotland is 11·76.

During the last hundred years or so, there has been some consolidation and enlargement of holdings, but the high proportion of the land occupied by owners seems always to have been a feature in the county. It is not, perhaps, surprising that this was the cause of delay in undertaking agricultural improvements. "The whole county," says the *New Statistical Account*, "till a very recent period was wild and barren, which circumstance has been attributed to the local peculiarity of the district being divided into small farms, almost every single farm being a separate property and generally possessed by its owner." What is, however, remarkable is that the late commencement and slow progress of improvement should be ascribed not to the owners' lack of capital but to their lack of energy. It has generally been held that one of the greatest advantages of a system of peasant

The Well at Scotlandwell

proprietors is that the farmers have the incentive to work with greater assiduity and diligence as owners than they would as tenants; yet of Portmoak parish we read: "The chief obstacle in the way of farming has been the practice of small proprietors working along with their servants. As they do not feel themselves called upon to work hard, the servants imitate their example." The

Agrarian Revolution, however, when it did reach Kinross, altered completely the appearance of the county. Before, only a small proportion of the land was in regular cultivation. This was called "infield"; the "outfield" was the surrounding land in its natural state or just broken up and under no regular cultivation. Much of the country was wild unsheltered moor, the land was not enclosed, and the houses of the cottars were placed irregularly in the open fields. Sour peat bogs were numerous and the undrained land held stagnant and injurious water. The great need was better drainage, which, for the southern portion of the county, was obtained by the widening and deepening (1811–1830) of the Pow of Aldie to form a complete main drain seven and a half miles in length. With what success, let the *New Statistical Account* tell: "The valley of the Gairney and the Pow is now completely changed and when viewed from the higher grounds forms a beautiful prospect especially to those who were acquainted with it in its former state." The lowering of the level of Loch Leven by the New Cut gave a greater fall and facilitated the work of drainage improvement. Gairney Moss, once 50 acres in extent, was converted into corn land or planted with trees; the peat bog to the south of Aldie was made to yield excellent crops; and, indeed, most of the peat mosses in the county were drained and put under the plough. They are now devoted to the cultivation of potatoes or oats, two crops both tolerant of the considerable amount of acid humus in such a soil.

With the improvements in drainage, the introduction

of a better rotation of crops and the invention and utilisa-
tion of farm machinery, farming became more profitable,
and, by the middle of the nineteenth century, almost
three-fourths of the total area of the county was under
cultivation. This proportion has now dropped to two-
thirds, and another notable change is the reversion of
arable land to permanent pasture, though neither of these
changes is peculiar to Kinross. There are only two
counties in Scotland (Fife and Linlithgow) which have
a greater proportion of their area under cultivation. The
fluctuations in the amount and direction of agricultural
effort are shown in the following table:

*Area of Cultivated Land expressed as a Percentage
of Total Area*

1854	1881	1911
73	63	66

*Area under various Crops or Grass expressed as a
Percentage of Total Area under Cultivation*

	1854	1881	1911
Wheat	1	—	—
Barley	8	4	—
Oats	20	18	18
Potatoes	2	3	2
Turnips	10	8	7
Rotation Grasses	37	36	34
Permanent Pasture	16	29	36

The year 1854 was a good year for farmers, the
Crimean War having stopped supplies from Southern
Russia. This increased the price of grain but did not,

indeed, affect the area under cultivation for that year, though the high profits naturally led to some increase in the area devoted to crops during the next few years. The year 1881 may be taken as the end of a period of depression; there had been seven very bad seasons between 1872 and 1881, and farmers had sustained very heavy losses.

Town Hall, Milnathort

Where the boulder clay is covered with a deposit of sand and gravel, as is usual in the plain of Kinross, the resulting soil is light and sharp and requires a liberal use of lime. The proportion of grass on every farm is large, a good deal of land is kept permanently in grass parks, and increasing attention is given to dairy farming. Only 90 acres are under wheat, an exhausting crop and, owing to the somewhat severe spring weather, a precarious one. The only grain market in the county is at Milnathort.

Stock raising is as important as agriculture. Thirty years ago more horses were reared in Kinross in proportion to its size than in any other Scottish county. The cattle number 6655, mostly shorthorns and Ayrshires or crosses bred from these. The rapid increase of the population in the mining districts of south-west Fife has stimulated the dairying industry, much of the milk being sent by railway to Dunfermline and other centres. With the reversion of arable land to pasture, the number of sheep has steadily increased; these, chiefly black-faced or Leicesters, now total 34,040. In former times there were few families among the labouring population that did not keep one pig or more, but the number of pigs in the county has now declined to 850.

Number of Live Stock in the County per 1000 *acres*

	1854	1881	1911
Horses	28	21	22
Cattle	147	111	127
Sheep	421	532	649
Pigs	21	10	16

The county is not well wooded, the area under woods and plantations being, in 1905, only 2922 acres. Curiously enough, the most extensive plantations are on the Blairadam estate, where, in 1733 before the improvements started, "the estate was little better than a wild unsheltered moor, the bleakness of which was increased rather than relieved by one solitary tree."

8. Industries and Manufactures.

Apart from farming and coal-mining, the county has practically no special industries of any magnitude. The ordinary work is to provide for common every-day needs —food, shelter, clothing, and comfort; the men are engaged in the preparation and sale of provisions, in house-building and decorating, in the administrative service or in the professions; the women are occupied in domestic service or dressmaking. It has already been mentioned that, to deal with the traffic that passes through the county, as many as 132 men are engaged in the railway service. Similarly the figures for other branches connected with the transport service are comparatively high, there being 35 blacksmiths in the county while 68 men find work in the conveyance of passengers or goods by road.

In the past several attempts were made to establish certain manufactures. In the little village of Kinnesswood the manufacture of vellum and parchment was established, it is said, by the monks of St Serf. It certainly flourished there for a long time and, from the middle of the seventeenth century, used to supply the Register House in Edinburgh. The last Census, however, returns only six persons as engaged in the manufacture of paper. The town of Kinross during the eighteenth century was famous for its cutlery. This manufacture was fostered by the excellence of the grindstone obtained from a quarry whose site can still be recognised by a scar on the side of the Bishop Hill. Towards the end of the eighteenth century, the cutlery industry was killed by the competition of Sheffield, and the people were devoting themselves to the manufacture of linen, especially Silesia linens. By 1839 this manufacture, unable to compete with Dunfermline, had collapsed. More attention was now given to the weaving of cotton, which had been started about 1809. The *New Statistical Account* (1839)

Bishop Hill, showing scar which marks the old
grindstone quarry

says: "A great many people are employed in weaving cotton, chiefly by the manufacturers of Glasgow, and, within the last twelve months, two or three companies belonging to Kinross and Milnathort have set agoing the manufacture of tartan shawls and plaids which hitherto appears to have met with success." This success, how-ever, did not last for many years. Later a revival took

place; and at the present time 129 persons are engaged
in the woollen industry, the mills and factories being at
Kinross and Milnathort, while 108 persons work in the
linen manufacture at Kinross; but the cotton industry
has been completely abandoned.

9. Mines and Minerals.

Next to agriculture, mining is the most important
industry in Kinross. The 364 people engaged in this
industry all live in the extreme south, for the coal of
Kinross is obtained from the extension of four of the Fife
coalfields. An outlying basin of the Dunfermline coalfield has a breadth of more than a mile on the Blairadam
estate. The coal has been worked round the margin of
a dolerite sill, especially to the south-east of Blairadam
House, where two seams are found, ten feet and four
feet thick respectively. Coal undoubtedly underlies the
sill but it may prove to be unworkable because of interfer-
ence with the seams by intrusive igneous material. The
results of borings, however, have been considered sufficiently
satisfactory to justify a mineral line being laid from Kelty
station to the estate. The extreme northern border of
the Kelty coalfield, and further east the Capeldrae field,
cross the border of Kinross. The Capeldrae seams have
been greatly broken up by faults and igneous intrusions,
which have rendered mining costly and troublesome.
The fourth field, that of Kinglassie, reaches North
Bogside, and is also connected by a mineral line with

Kelty station. Separate figures cannot be obtained for the Kinross-shire output of coal, since the returns are included with those of Fifeshire; but it is certainly increasing and the Census Report for 1911 shows that 313 miners reside within the county as compared with 186 in 1901. No other mining or quarrying is carried on; but sand and gravel are easily obtained in the plains, and the value of the 12,458 tons dug in 1911 was £2212.

10. Fishing.

Loch Leven has long been famous for its fish and Kinross is a great centre for anglers. Excellent food abounds in the lake, which is especially noted for trout and perch. At one time char (*Salmo alpinus*) was frequently caught, both Sibbald and the author of the *Old Statistical Account* mention that fish as being abundant, but the last recorded capture was in 1837. It is, however, the Loch Leven trout (*S. levenensis*) that has made the loch most famous. It is still disputed whether this fish is a distinct species or merely a superior variety of the common trout (*S. fario*) resulting from the better quality and greater quantity of the food supply. The chief differences between *S. levenensis* and *S. fario* are in the number of the caecal appendages, in the markings of the body, and in the colour of the flesh. The caecal appendages in *levenensis* vary from 60 to 80, in *fario* they never exceed 50. Hence the Loch Leven trout has been named

S. caecifer. The upper parts of the body of the Loch
Leven fish are usually greenish-olive, sometimes brownish,
and are marked with a great number of round or
X-shaped dark brown spots. In *fario* again red spots
are general and numerous. So also the adipose fin in
levenensis is never tipped with red, which is universally
the case with *fario*. The flesh of the latter is whitish,
that of *levenensis* is a deep pink. The Loch Leven trout
is also much more slender in form and the hinder part is
more tapering, the maxillary is much narrower and feebler,
the pectoral fins more pointed, the caudal more deeply
incised and with the lobes more pointed than in the same
size of *fario*. Dr Francis Day, however, considered the
Loch Leven trout to be a variety, not of the common
trout, but of the salmon, or sea, trout (*S. trutta*). The
maximum length of *S. levenensis* is 20 inches from eye to
fork. It is a non-migratory species, peculiar to Loch
Leven and a few other lochs of Southern Scotland and
Northern England; but has been artificially introduced
into many lakes not only in this country but also throughout the world—even as far away as Tasmania.

The fishings of the loch were let to tacksmen till
1876, when the Loch Leven Angling Association was
formed. The rent in 1780 was only about £20, by
1838 it had risen to £204; and at the present time the
lessees, the Tay Salmon Fisheries Company, Ltd., pay
£1200 a year with the right to put 36 boats on the lake.
The fishing season extends from the 31st March to the
7th September. Only rod fishing is permitted; most
anglers use the fly, but some try trolling with the minnow.

The catch for the season of 1910, a record one, was 41,064 trout of an aggregate weight of 29,200 lbs. These figures may be compared with the average annual catch for the years 1895–99 : 16,746 trout of 11,421 lbs. weight. The heaviest trout ever caught in the lake

An Angling Club starting from Loch Leven pier

scaled nearly 10 lbs., but the average weight is about 1 lb. Some years ago—especially in 1902 and 1903—great trouble was caused by Canadian weed, which not only interfered with the angling but also rendered rowing very difficult in the shallow waters.

11. History.

The history of the county has always centred round
Loch Leven. It seems probable that, in prehistoric times,

Castle Island, Loch Leven

the Castle Island was the site of a crannog, or stockaded
lake-dwelling; and the stone work of another has been
located near the shore of the little Kinross peninsula,
where a beacon can be seen marking the position. These
crannogs, peculiar to Celtic countries, were constructed

on islands or even shallows in the lochs. In the latter
case, the site was built up by stones, brushwood, or other
material, to make occupation possible. The occupied
area was circumscribed by one or more lines of wooden
piles, and often connected with the shore by a causeway.
An ancient causeway does connect Castle Island with
the shore; and it is said to be possible, in a dry season,
for a man to wade along it. Security from attack was
one essential requirement for prehistoric settlements; and
hence, besides the crannogs in the lakes, evidences of
other defensive works are found to-day in the mounds
encircling the crests of some of our hills. There are, for
example, traces of an ancient fort on the south-west
shoulder of Benarty; and of another on the top of
Dumglow, the highest of the Cleish Hills. Besides forts,
camps, and crannogs, the weapons of the early inhabitants
and their ornaments are sometimes unearthed, while the
sepulchral monuments, raised over their dead, are scattered
throughout the country. There are two "standing
stones" adjacent to the farm of Orwell near the north
end of Loch Leven; urns filled with burnt bones have
been discovered at Holeton, another farm in the same
parish; while, till the stones were removed to build field
dykes, a great cairn stood on Cairn-a-vain, a hill north-
north-west from Milnathort. Inside this cairn was discovered, about the year 1810, a rude stone cist containing
an urn full of charred bones and charcoal. Agricola's
legions most likely visited the shire; and the discovery
in 1857 of a hoard of 700 Roman coins near the town
of Kinross may, unless the coins represent a collection

subsequent to the Roman withdrawal, be an indication of the passage of a Roman army.

Since the prehistoric times, however, the history of the county has mainly been the history of the two largest islands in Loch Leven. These islands are both at some distance from the shore and the people who lived on them would be isolated, quiet, and safe. The one island became the seat of a famous monastery; the other the site of a castle, to the Constable of which important State prisoners were committed for safe keeping. The ecclesiastical settlement on St Serf's Island is said to have been founded by Brude V, the last of the Pictish kings. Servanus was the first superior or prior, but it is impossible to discover whether or not this Servanus is the same St Serf of Culross whose labours converted the Picts. The Register of St Andrews says of this foundation: "Brude, filius de Ergard, Pictorum rex dedit insulam de Loch Levin, Deo Omnipotenti, Sancto Servano, et Keledeis heremetis ibi commorantibus et Deo servientibus." By the tenth century the name of Culdees had become established as that of the order of ecclesiastics to whom the priory of St Serf's Island, as well as numerous other establishments in Scotland and Ireland, belonged. Their monastic rules were not, however, of a very strict order and each monastery appears to have been practically independent. For a time they were in great favour with the monarchs; Macbeth and his queen Gruoch granted to the hermits of St Serf's Island—the oldest Culdee establishment in Scotland—the lands of Kirkness and Portmoak; and what is now the old burial ground of Portmoak is the site of a monastery which was a secondary establishment of the island community, whose priors and canons, indeed, often resided at Kinnesswood. Other bequests followed, but later came the disputes as to rule and discipline between the Culdees of the ancient Celtic Church and the clergy of the Romish church, first introduced by Malcolm and Margaret. Towards the close of the tenth century the priory of

The Chapel, St Serf's Island

Loch Leven and its possessions were made over to the Bishop of St Andrews on condition that he should supply the monks with food and raiment; and, in 1145, David I issued a declaration bestowing the island of Loch Leven on the Canons Regular of St Andrews that they might establish a Canonical Order there, and decreeing, further-more, that all the Culdees who should not consent to live as Regulars should be expelled. The Canons of

St Andrews held the island till the Reformation, when it passed into the possession of the Earl of Morton. It is noteworthy that the old Culdee community at Loch Leven gives us the oldest Scottish library catalogue, which is preserved in the Register of St Andrews. Remains of the old chapel on St Serf's Island can still be seen. The old building is 30 feet by 20 feet, and the door on the south side is 8 feet high. The ruins were explored in 1877 and two skeletons were discovered, supposed to be those of St Ronan and Patrick Graham. Graham was Bishop of St Andrews at the time when that see was erected into an archbishopric, and was buried on St Serf's Island, 1484.

The Castle Island appears to have been inhabited from the remotest period of antiquity. Congal, the son of the Pictish king Dongart, is said to have built a stronghold on the island, and a castle there was certainly a royal residence during the first half of the thirteenth century. This castle was twice besieged by the English, in 1301 and in 1334. On the second occasion the English leader, Sir John de Strivelin, encamped his force on the consecrated ground of Kinross kirkyard, a proceeding which Wyntoun, who tells the story of the siege, strongly reprobated. From Lent till summer the attacking force strove to reach the enemy, but the Scots, under the able command of Sir Alan de Vipont, repulsed their efforts and the siege was ultimately abandoned. Buchanan, followed by Hailes, says that the English attempted to submerge the castle by damming the issue of the water at the south-east end of Loch Leven, but the garrison,

in the absence of the English leaders at the festival of
St Margaret in Dunfermline, sallied forth and broke
down the dam, the escaping waters sweeping away the
English encamped on the plain below.   Buchanan's story
is, however, improbable, partly from the difficulty of the
undertaking and partly because a flood which would have
submerged the Castle Island would also have swamped
St Serf's.   Had this really occurred, it would most
certainly have been mentioned by Wyntoun—in later
years, Prior of the establishment there—on whom the
sacrilege of such a proceeding would have made a great
impression.   Yet Wyntoun has no word at all about any
attempt to submerge the castle.

   Apart from this siege, the most interesting incidents
in the history of the castle are connected with the long
succession of State prisoners who were immured within
its walls.   The first of these seems to have been John
of Lorn, Lord of the Isles.   In 1363 David II sent as
prisoners to the castle his nephew Robert the Steward
of Scotland and Robert's son, the "Wolf of Badenoch."
The next prisoner of note was Archibald, fifth Earl of
Douglas, confined here by the orders of James I in 1429.
Patrick Graham, the first Archbishop of St Andrews,
spent the last few years of his noble and beneficent life
imprisoned within the castle walls.   But the most pathetic
figure ever ferried across to the island prison is that of
Mary Queen of Scots.   It was on the 16th day of June,
1567, after Bothwell's escape and the surrender of the
royal forces to the confederate lords at Carberry Hill, that
she was brought, through Edinburgh, as a prisoner to

Lochleven Castle. The laird of the castle was Sir William Douglas. His mother was also there, who before her marriage had been Lady Margaret Erskine. Other members of the Douglas family in the castle were Sir William's brother George, his wife and sisters, and Willie Douglas, the sixteen year old page of Lady Margaret. As the mother by James V of James Stewart, Lady Margaret had cherished the hope of seeing her son crowned king and she was, in consequence, the bitter enemy of Mary. The provision made for the poor queen was at first shamefully inadequate until her own fascination or Moray's orders won for her more fitting treatment. Mary, moreover, was harassed by visits from the lords who came to extort from her consent either to a divorce or an abdication. The story of her imprisonment and of her attempts at escape has often been told; how, after one plan had been foiled, she, changing clothes with a laundress, entered the boat only to be discovered by the boatmen midway to Kinross and brought back to the castle; how George Douglas, who had assisted this attempt, was then expelled from the castle but remained in the vicinity of Kinross to scheme and work on the queen's behalf; and how, by the help of Willie Douglas, the "Roland Graeme" of Scott's *Abbot*, she eventually succeeded in escaping. The chance came on the evening of Sunday, the 2nd May, 1568. Waiting on Sir William at supper, Willie managed to remove the keys and to slip unnoticed with them from the room. While the laird lingered over his wine, Mary, Willie, and one of Mary's maids ran down the stairs and, locking the gate behind

them, gained the boats. The boy had taken the pre-
caution to leave but one of these with the chain unjammed,
and in this the fugitives rowed to where George Douglas,
Lord Seton and a troop of horse were waiting on the
shore.

Soon after Mary became a prisoner in England,
the Earl of Northumberland plotted to rescue her, and
failed. He fled to Scotland, where at Elizabeth's request
he was imprisoned. For three years he remained a
prisoner in Lochleven Castle, till in 1572 he was handed
over to the English, and beheaded at York. The last
State prisoner in the castle was Robert Pitcairn, Arch-
dean of St Andrews, who was implicated in the Raid
of Ruthven, and was imprisoned for a short time in
1583.

In later days the only event in Kinross-shire of
national importance is connected with ecclesiastical affairs.
The Patronage Act of 1712 caused much trouble in the
Church of Scotland. In 1731 Thomas Erskine, minister
of Portmoak for nearly 30 years, was translated to
Stirling, and there he preached a notable sermon advocat-
ing the right of congregations to choose their ministers.
Erskine and three other ministers were suspended by
the General Assembly, and in December, 1733, they
met at Gairney Bridge to constitute themselves the
Associate Presbytery. This was the origin of the
Secession Church, which—in spite of disputes and
divisions into Burghers and Anti-Burghers, Auld Lichts
and New Lichts—spread and prospered till in 1847
it was able to contribute 384 of the 497 congregations

which then combined as the United Presbyterian Church.

Monument to the founders of the Secession Church at Gairney Bridge

12. Architecture[1].

Lochleven Castle is a good example of the type of building erected during the fourteenth century. It is a plain rectangular structure with walls of great thickness confining an area of some 40 square yards. The floors are vaulted, and the roof is crowned by a parapet resting on corbels but without machicolations.

Queen Mary's Tower, Lochleven Castle

There are not, however, bartizans at all the corners, for it was, presumably, not thought necessary to defend by such a turret the corner nearest the interior of the courtyard. A peculiar feature is the position of the entrance door, which, instead of being on the first, is placed on the second floor, and thus it is possible

[1] See p. 55.

to reach the first floor only by descending the interior staircase. A parapet walk runs round the wall of enceinte; and, in the south-east corner of this wall, is a round tower, built, probably by Sir Robert Douglas, about 1541. When Mary was on the island, it contained a more extensive range of buildings than those whose remains still exist; and there is evidence that the usual development by building round the inside of the wall of enceinte had taken place.

Burleigh Castle, close to Milnathort, probably dates from the latter half of the fifteenth century. Formerly there appear to have been a keep and a wall of enceinte with buildings added thereto, but nothing now remains except the ruins of the old keep and a circular tower or gatehouse connected with the keep by a portion of the west wall. In the angle between the tower and the wall is placed a turret containing the stairs leading to the upper floors. The circular tower is the most interesting portion of the building. The basement walls of this gatehouse are pierced by large horizontal embrasures; and the room on the first floor, resting on the vaulted basement roof, is square internally. The second floor is square also externally, being boldly corbelled out. The gables are finished with flat skews—a very unusual feature—instead of the customary corbie steps, and both of the skew-puts bear a device. On the one facing the modern road is the red rose of the Balfours; on the other is the date of the tower, 1582, above a shield carrying the Balfour arms and the initials M. B. and I. B. These stand for Margaret Balfour, the heiress of the Balfours (to whom the Burleigh property had been granted by James II in 1446) and for Sir James Balfour of Mountquhane, her husband. Their son, Sir Michael Balfour, after a successful career in the diplomatic service, was raised to the peerage by James VI in 1606 as Lord Balfour of Burleigh. Later the peerage was in abeyance for nearly 150 years until the Bruces of Kennet in 1868

Burleigh Castle, Milnathort

established their claim to the title. Both Lochleven and Burleigh Castles are now the property of Sir Basil Montgomery. Another keep of about the same period as Burleigh is Arnot Tower near Auchmuirbridge. In spite of its somewhat superior masonry work, the building is in a very dilapidated state. It was for more than 600 years the family seat of the Arnots of that ilk.

On the northern slopes of the Cleish Hills are situated

two castles, Cleish and Dowhill, dating from the sixteenth century. The former, built on the L plan, after falling into such a state of disrepair that the roof had entirely collapsed, was restored and made habitable by the proprietor, Mr Harry Young, in 1845. As a result, it has been greatly altered; the thick walls have been thinned, the entrances changed, the windows enlarged, the vaulted floor removed, but the external features were preserved as far as possible, and, in particular, there still stands, on the south side, the great gable which rises with three broad offsets to a height of some 70 feet. The estate came into the possession of Sir James Colville of Ochiltree in the early part of the sixteenth century; and his descendant, Robert Colville of Cleish, was created Lord Colville of Ochiltree by Charles II in 1651. The line became extinct on the death of the third lord in 1723. Dowhill Castle also fell into disrepair and, since it was used for a time as a quarry, it is now in such a ruinous condition that we can only say that it appears to have been built on the Z plan. The ground floor and a portion of the side walls of the first floor remain, but the chief feature of interest in the castle is the superior finish of the interior stonework.

Tullibole Castle, one mile east of the Crook of Devon, is an old baronial mansion of the seventeenth century. As in that quieter time there was less need for defensive arrangements, the embattled roof characteristic of the earlier castles has almost disappeared and the interior accommodation is much better, notably in the provision of staircases. The only part of the roof adapted for defence is a small battlement, supported on bold corbels and overhanging the doorway of the square tower at the south-east corner. Above this doorway is a large panel bearing the date, 1608, the initials of John Haliday and his wife Helen Oliphant, their arms and a couple of mottoes. Standing on a hill overlooking the Pow Burn,

Tullibole Castle

about a mile or so to the south of Tullibole Castle, is the older castle of Aldie. Erected probably before 1500, the keep has the usual defensive features. The building added on the south side is later.

Of the many fine mansions in the shire it is possible here to mention only two. Kinross House was built by the celebrated architect Sir William Bruce in 1685. It is an imposing building and was so greatly admired that it became the model for many later erections throughout the country. Sir Robert Sibbald, in whose lifetime the house was built, writes of it as "a stately building... which for situation, contrivance, prospects, avenues, courts, gardens, gravel-walks and terraces, and all hortulane ornaments, parks and planting, is surpassed by few in this country." William Adam, a pupil of Sir William Bruce, built for himself the mansion of Blairadam in the south of the county. Apart from its architectural features, Blairadam House is also of interest as being the place where Sir Walter Scott and the other members of the Blairadam Club foregathered to spend the week-ends with their host, the grandson of Bruce's pupil.

There is little of interest in the ecclesiastical architecture of the county. The small ruin remaining on St Serf's Island is scarcely worth the trouble of a visit, and no vestige remains either of the priory of Portmoak or of the hospital at Scotlandwell. With the exception of Orwell church, the existing parish churches were all built during the first 40 years of the nineteenth century and are mostly Gothic in style. Orwell is only an apparent exception, for, though the building at Milnathort was erected in 1729, it was entirely renovated at a later date. In the old burial-ground of Tullibole, just to the south of the foundations of the church demolished in 1729, there lay till recently a sculptured stone, now in the Antiquarian Museum, Edinburgh. Over four feet high, about a foot and a half broad, it consists of close-grained freestone, and is sculptured on both sides and on the edges. One side has a man on foot ; another on horseback with hound and other beast ; two men

Sculptured Stone found at Tullibole

wrestling ; two serpents with looped tails, face to face. The other side had an elaborately ornamented cross, which is now much obliterated.

13. Distribution of Population. Communications.

Kinross-shire has the smallest, though not the least dense, population in Scotland. The total population n 1911 was 7528 persons, of whom one-third resided n the only burgh in the county, Kinross. Farming

Curling on Loch Leven

being the only important industry, the population is scattered in small hamlets and farmsteadings throughout the plain. Such a district needs a centre where farm produce can be marketed and where the farmers can purchase commodities. This market town will arise at the point most conveniently accessible for the whole district. In the case of Kinross-shire, the most convenient situation would naturally be near the centre of the basin, where, in fact, we do find the town of Kinross. The town appears to owe its size and importance to three chief factors : (1) it is the natural centre and, therefore, the market town of the whole shire ; (2) it is on the Great North Road, roughly about half-way between Edinburgh and Perth, and was an important stopping-place in the old coaching days ; (3) it is on the shore of Loch Leven and has become a centre for skaters and curlers in the winter and for anglers in the fishing season. In past ages its position at the point nearest the Castle Island lent it additional importance. The only other urban centre is Milnathort. It, too, is situated on the Great North Road, but at the point of conjunction of roads from Cupar, Leslie, and Dollar.

Another result of the almost exclusive devotion to farming is that the population keeps steadily at a fairly constant figure and shows little signs of increase. Allowing for the alteration of boundaries, the number of inhabitants is now but 5·7 °/₀ above the number in 1801, a rate of growth which may be contrasted with the 153·4 °/₀ increase of the population of Clackmannan, or the 195·9 °/₀ increase of the population of the whole of Scotland since the same date. The density of population per square mile in the various parishes is as follows : Kinross, 273 for the whole parish, 43 for the extra burghal portion only ; Orwell, 96 ; Portmoak, 55 ; Cleish, 54 ; Fossoway, 43.

As regards railway communication, Kinross is well served. The North British Railway from the Forth

Bridge enters the county between Benarty and the
Cleish Hills and runs north to Kinross, where it is joined
by the Devon valley branch from Alloa.   Proceeding

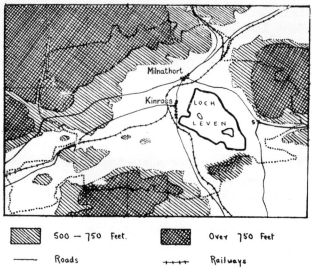

Sketch-map of Kinross-shire, showing the influence of the
physical features on the lines of communication

round the north-west end of Loch Leven, a line diverges
at Mawcarse Junction to pass down the Eden valley,
while the main line continues northwards through
Glen Farg to Perth.

14. Roll of Honour.

Of famous names in the history of Kinross-shire
we have already mentioned St Serf; Mary Queen of
Scots; her custodian, Sir William Douglas, who became
the seventh Earl of Morton; and the founder of the
Secession Church, Thomas Erskine, who as minister of
Portmoak drew hearers from far and near.

In the seventeenth century and the eighteenth, the
county produced a succession of famous architects.
William Bruce, Monk's messenger to negotiate with
Charles II in Holland, became a baronet in 1668,
King's Surveyor and Master of Works in 1671, and
later hereditary sheriff of Kinross. Besides his own house
of Kinross, he restored Holyrood for the king, built
Harden House in Teviotdale, and the Merchants' Hall
at Glasgow, and designed Hopetoun House in Linlithgow.

On Sir William's death at a very great age in 1710,
his pupil, William Adam, a native of Kinross, became
Surveyor of the King's Works in Scotland, and, in later
life, published *Vitruvius Scoticus*, a selection of his designs
together with those of his contemporaries. He died in 1748
and his two sons, Robert and James, removed to London
and became two of the most famous British architects of
the eighteenth century. These two brothers designed
the Adelphi buildings in London; a fact which accounts
for the name.

Another noted name of the eighteenth century is
that of Michael Bruce, born at Kinnesswood in 1746.
The son of a weaver, he studied at Edinburgh University,

intending to enter the ministry of the Secession Church. He was schoolmaster at Gairney Bridge, and then in Clackmannanshire at Forest Mill. Bruce was in ill-health and rapidly grew worse in his unwholesome surroundings.

Sir William Bruce

The floor of the schoolroom was the bare earth, which in rainy weather degenerated into a quagmire of wet mud; and, though the parents of his scholars did their best by laying a row of stones from the door to the master's desk,

Michael's health completely broke down and he returned home to Kinnesswood to die of consumption at the age of 21. Bruce was a poet. His longest poem is *Lochleven*, the following extract from which, though in the conventional eighteenth-century style, is vividly descriptive of the actual scene :

Michael Bruce's Cottage at Kinnesswood

" Between two mountains, whose o'erwhelming tops,
    In their swift course, arrest the bellying clouds,
    A pleasant valley lies. Upon the fourth,
    A narrow op'ning parts the craggy hills,
    Through which the lake, that beautifies the vale,
    Pours out its ample waters. Spreading on,
    And wid'ning by degrees, it stretches north
    To the high Ochil, from whose snowy top
    The streams that feed the lake flow thund'ring down."

9—5

Better known are Bruce's paraphrases from scripture, and his short lyrics. A collection of his works was published in 1770 by Rev. John Logan, a former fellow-student, who, however, reserved several poems to issue later in a volume of his own. This makes it doubtful if the *Ode to the Cuckoo*, a really fine poem, is Bruce's or not.

A much earlier writer is Andrew of Wyntoun, born about 1350. A Canon Regular of St Andrews, he was made Prior of the Monastery on St Serf's Inch. There he compiled his *Orygynale Cronykil*—original in the sense of beginning at the "origin," the creation of the world. It is only in the Scottish parts from Malcolm Canmore that Wyntoun is historically of much value. His chronicle is in the octo-syllabic metre, but as poetry it is worthless. Wyntoun gives us the original version of Macbeth's interview with the weird sisters, which as embellished by Hector Boece and repeated by Holinshed forms the groundwork of Shakespeare's *Macbeth*.

15. THE CHIEF TOWNS AND VILLAGES OF KINROSS-SHIRE.

(The figures in brackets after each name give the population in 1911, and those at the end of each section are references to pages in the text.)

Cleish on the Gairney Water, is the only village of any size in the parish of Cleish. The wonderful adventures of Squire Meldrum of Cleish are told by Sir David Lyndsay. (p. 128.)

Crook of Devon, a village situated where the Devon changes the direction of its course from south-east to south-west. Formerly a burgh of barony and a place of some note for its cattle fairs, it now derives a slight importance from its mills, and its position as a convenient stopping place for visitors to the picturesque river defile. (pp. 25, 29, 128.)

Kinnesswood, a village at the western foot of the Bishop Hill. Like the other villages (e.g. Easter and Wester Balgedie) of Portmoak parish, it has probably grown from a fishing settlement; but, with the lowering of the level of the loch, it is now some distance from the actual shore. Till the emigration of the Birrell family it had been long famous for the manufacture of vellum and parchment. Portmoak was the birthplace of John Douglas, first "tulchan" Archbishop of St Andrews. (pp. 119, 135.)

Kinross (2618), created a burgh of barony by the Regent Morton, and now the only police burgh in the county, is

Rumbling Bridge

the county town. It is also the official headquarters of the Scottish Amateur Skating Association and the National Angling Association. (pp. 84, 96, 97, 110, 112, 113, 132, 134.)

Milnathort (1178), in Orwell parish, a market town, one and a half miles north of Kinross. (pp. 84, 96, 109, 112, 133.)

Rumbling Bridge is a famous centre for visitors to the gorges and falls of the Devon river, here spanned by two bridges— the older dating from 1713, the modern not yet a century old. The defile at this spot is about 120 feet deep. Above the bridges is the Devil's Mill, below are the falls of the Caldron Linn.

Scotlandwell, an ancient village in Portmoak parish at the southern foot of the Bishop Hill. The name was taken from the " Fontes Scotiae," famous springs, one of which, now protected by a wooden structure, rises just to the west of the main street. In 1238 William Malvoisin, Bishop of St Andrews, founded a hospital for the Mathurin or Red Friars, on a site now marked by the old burial-ground at the south-east extremity of the village. (pp. 98, 105, 130.)

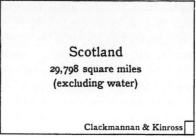

Fig. 1. Comparative areas of the counties of Clackmannan
(54 sq. miles) and Kinross (58 sq. miles) and all Scotland

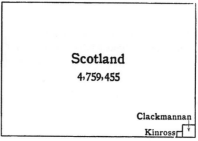

Fig. 2. Comparison in Population of Clackmannanshire
(31,121) and Kinross-shire (7528) and all Scotland

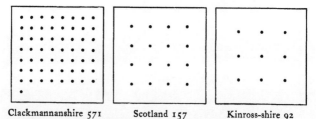

Fig. 3. Comparative Density of Population to the
square mile in 1911

(Each dot represents ten persons)

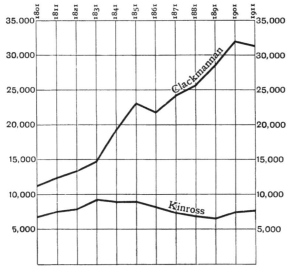

Fig. 4. Graph showing Growth of Population in the
counties of Clackmannan and Kinross since 1801

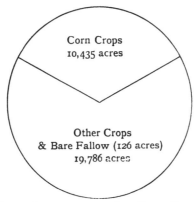

Fig. 5. Proportionate area under Corn Crops compared
with that of other cultivated land in the counties of
Clackmannan and Kinross in 1912

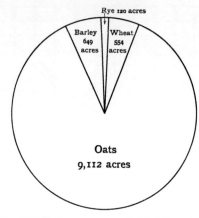

Fig. 6. Proportionate area of chief Cereals in the counties of Clackmannan and Kinross in 1912

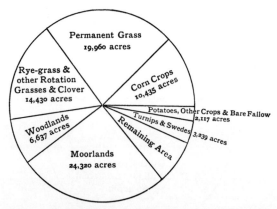

Fig. 7. Proportionate areas of land in the counties of Clackmannan and Kinross in 1912

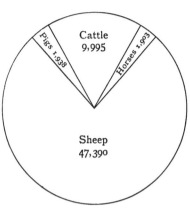

Fig. 8. Proportionate numbers of Live Stock in the
counties of Clackmannan and Kinross in 1912

www.ingramcontent.com/pod-product-compliance
Ingram Content Group UK Ltd.
Pitfield, Milton Keynes, MK11 3LW, UK
UKHW042145280225
455719UK00001B/119